Climate Change: A Natural Hazard

William Kininmonth

phys

.18H22HH3

ISBN 0 906522 26 9
Multi-Science Publishing Co. Ltd.
5 Wates Way, Brentwood, Essex CM15 9TB, UK

The Author

William Kininmonth has a career in meteorological science and policy spanning more than 40 years. For more than a decade (1986–1998) he headed Australia's National Climate Centre with responsibilities for managing the data archives, monitoring Australia's changing climate and advising the Australian government on the extent and severity of climate extremes, including the recurring drought episodes of the 1990s.

William Kininmonth has extensive knowledge of global climatology, the climate system and the impacts of climate extremes developed through more than two decades associated with the World Meteorological Organization. He was Australia's delegate to the WMO Commission for Climatology and more recently has been a consultant for implementation of its programs. He coordinated the scientific and technical review for the United Nations Task Force on El Niño following the disastrous 1997–1998 event, has participated in WMO expert working groups and carried out training workshops for climatologists from Africa, Asia and the Pacific.

As a member of Australia's delegations to the Second World Climate Conference (1990) and the subsequent intergovernmental negotiations for the Framework Convention on Climate Change (1991–1992), William Kininmonth had a close association with the early developments of the climate change debate. His suspicions that the science and predictions of anthropogenic global warming had extended beyond sound theory and evidence were crystallised following the release of the 2001 Third Assessment Report of the Intergovernmental Panel on Climate Change.

Table of Contents

Table of Figures

A Rush to Judgement

"It is now believed that in the first half of the next [21st] century a rise in global mean temperature could occur which is greater than any in man's history."[1]

The road to Kyoto

According to Mark Twain, weather is what everybody complains about but what nobody does anything about. However in October 1985, at an international meeting in Villach, Austria convened by United Nations agencies, a group of scientists decided it was time for the world to take action. The meeting concluded that there was a need to combat the perceived danger of global warming that would result from increasing concentrations of so-called greenhouse gases in the atmosphere. The greenhouse gas concentrations, particularly those of carbon dioxide (a product of burning coal, oil and other fossil fuels) are increasing as a direct consequence of a range of human activities, including industrialisation, land-use practices and transport. The Villach Conference Statement was a dramatic prophecy that launched the threat of climate change on a largely unsuspecting public.

The Villach Statement and its threat of global warming became an international rallying cry for action to curb emissions of greenhouse gases to the atmosphere. Around the world a diverse range of interest groups, especially across the environment movement, cooperated to raise public awareness of the greenhouse climate change threat. A series of government sponsored national and international conferences

---

1   Statement issued following the October 1985 United Nations co-sponsored conference at Villach, Austria convened for the purpose of making an "international assessment of the role of carbon dioxide and other greenhouse gases in climate variations and associated impacts."

of invited experts were widely reported in the media and ensured a raised public recognition of the issue. In addition, electronic and printed information material were prepared by national and international interest groups and distributed to the media and made available through educational channels.

So successful was the awareness-raising campaign that within 3 years the United Nations, through its agencies UNEP[2] and WMO[3], had established an intergovernmental mechanism to address anthropogenic[4] climate change. The Intergovernmental Panel on Climate Change (or IPCC) was charged with:

(i) "Assessing the scientific information that is related to the various components of the climate change issue, such as emissions of major greenhouse gases and modification of the earth's radiation balance resulting therefrom, and that needed to enable the environmental and socioeconomic consequences of climate change to be evaluated.

(ii) Formulating realistic response strategies for the management of the climate change issue".[5]

Three working groups were established to address the IPCC objectives. The tasks of Working Groups I, II and III were, respectively, to:

(i) Assess available scientific information on climate change.

(ii) Assess environmental and socio-economic impacts of climate change.

(iii) Formulate response strategies.

The IPCC clearly prejudged the findings of Working Group I and established the three working groups to operate concurrently. It was not a matter of identifying whether climate change was a realistic threat and then developing response strategies in a measured way. In any event, the confidence was not misplaced and the working group on science did confirm the Villach conclusions and found a serious

---

2   UNEP – The United Nations Environment Programme.
3   WMO – The World Meteorological Organization is a specialised agency of the UN.
4   'anthropogenic' is a term that has evolved in the climate change lexicon to become synonymous with a range of human activities, including industrialisation and land use changes that in some way change the concentrations of greenhouse gases in the atmosphere.
5   IPCC Working Group I Assessment Report, 1990

anthropogenic threat to the global climate. After a period of less than 18 months, in July 1990, the IPCC WG1 published their findings following an assessment of the available scientific literature. The principal findings of the report were:

(i) There is a greenhouse effect because a range of gases occurring naturally in the atmosphere, such as carbon dioxide, keep the earth's surface warmer than it would otherwise be.

(ii) The concentrations in the atmosphere of a range of greenhouse gases, including carbon dioxide, are increasing because of human activities.

(iii) The increasing concentrations of certain greenhouse gases in the atmosphere, such as carbon dioxide, will lead to global warming but neither its magnitude, timing nor regional characteristics could be determined.

Notwithstanding the last finding, the IPCC volunteered estimates of future global temperature rise (based on computer model projections) and sea level rise, which were essentially similar to those of the Villach Statement.

Not only was the IPCC review a rushed affair but so also was its 'validation' by the scientific community. Presented to the WMO sponsored Second World Climate Conference in mid 1990, the IPCC findings were endorsed in the concluding Scientific Statement. One wonders how many of the scientists at the Conference took the time to read the very lengthy and detailed volume. Some would have quickly reviewed those aspects pertaining to their own area of expertise and most would have perused the Summary for Policymakers. In any case, their views were largely inconsequential because the endorsement contained in the Scientific Statement was formulated by the organisers and presented for acceptance on the final day of the conference.

The United Nations General Assembly took up the challenge presented by the IPCC scientific assessment and the Statement of the Second World Climate Conference. An Intergovernmental Negotiating Committee was convened to develop a Framework Convention on Climate Change in time for the June 1992 Earth Summit at Rio de Janeiro. The Committee, open to all member countries of the United Nations, met on six occasions between February 1991 and May 1992 before finally reaching agreement. At the Earth Summit, representatives of more than 150 countries signed the United Nations Framework Convention on Climate Change (UNFCCC) that resulted from the negotiations. More countries have signed subsequently. The

Convention requires countries to take actions necessary for "stabili-sation of greenhouse gas concentrations in the atmosphere at a level that would prevent dangerous anthropogenic interference with the climate system".

Despite the perceived threat posed by anthropogenic global warming, the short period available for negotiations meant that agreement could not be reached on binding mechanisms for reducing greenhouse gas emissions and commitments that individual countries should make. Counterbalancing the global warming threat were the immediate economic and social costs that would be experienced by many countries if they took action to reduce greenhouse gas emissions. For example, developing countries have aspirations for living standards and lifestyles that other countries have achieved through industrialisation and mechanisation and wish to proceed on their respective developmental paths unhindered by controls over fossil fuel usage and land-use change. In contrast, those countries that have achieved developed economies are reliant on fossil fuels for energy and faced economic and developmental setbacks if fossil fuel usage is to be curtailed. Because of the complexity of the social and economic issues, and some uncertainty about the potential impacts of climate change, it was acknowledged that specific and binding actions by countries to reduce greenhouse gas emissions should be negotiated separately in Protocols to the Convention.

The IPCC continued its work and issued its Second Assessment Report in 1995. Contemporary experiments using computer models of the climate system and various natural and anthropogenic forcing functions pointed to anthropogenic signals that could be detected in the observed global warming pattern. The IPCC, in its Second Assessment Report, concluded that the balance of evidence suggested that a discernible human influence on global climate could be detected.

The public interest in the anthropogenic global warming issue and the perceived need for action did not abate. More than 10,000 people, made up mostly of non-government lobby groups and representatives of the world media, converged on Kyoto, Japan in December 1997 for the third meeting of the Conference of the Parties to the UNFCCC. They were there to witness government delegates negotiate a Protocol to stem the unconstrained emission of greenhouse gases into the atmosphere. The Protocol was expected to give teeth to the Convention, including binding commitments by industrialised countries to limit their emissions of greenhouse gases.

But the rush to judgement met stiff resistance at Kyoto. The negotiations were tense and the final text of the Protocol was agreed to by a number of developed countries with reservation.

Notwithstanding that a text was agreed at Kyoto, the Protocol has not yet come into force. The impediment is that a few industrialised countries reliant on fossil fuel energy resources and with increasing national greenhouse gas emissions, including the United States of America, Russia and Australia, have declined to ratify the Protocol. In order for the Protocol to come into force it is necessary for 55 countries that are Party to the Convention to ratify it, including sufficient Annex I Parties (developed countries) accounting for 55 percent of that group's carbon dioxide emissions in 1990. More than 55 countries have ratified the Protocol but with the failure of the USA and Russia to ratify it is the 55 percent criterion that has not been met.

Notwithstanding the stalled action on the political front, the scientific work has continued. The IPCC issued its Third Assessment Report in 2001, claiming that most of the warming of the previous 50 years has been caused by human activities. Confidence in the claim was drawn from the ability of newly developed computer models simulating the natural and anthropogenic radiation forcing (ie, changing solar radiation, greenhouse gas concentrations, etc) to reproduce the global temperature trend of the previous century. Also, a reconstruction of annual average northern hemisphere temperature from proxy data suggested that the 1990s was the warmest decade and that the rate and duration of warming over the 20th century was unprecedented over the previous millennium.

Flawed science

Much has been written about the economic costs of implementing the Kyoto Protocol, especially to those industrialised countries of the 'developed' world whose energy base is primarily fossil fuel. What has not been closely examined is the scientific construct adopted by IPCC that has led to its conclusions about anthropogenic global warming. The perception amongst policymakers is that the issue is beyond science and very much a balancing of projected impacts of anthropogenic climate change, including global warming and sea level rise, against the costs of reducing greenhouse gas emissions. That is, anthropogenic global warming has transcended to a political issue. However, what is not acknowledged is that the scientific basis for anthropogenic global warming is a flawed theoretical framework underpinned by inadequate computer models of the climate system.

The frailty of the scientific basis can be clearly demonstrated through an examination of the model of the earth's energy balance that is used to explain the concept of radiative forcing of the climate system. The construct of a global and annual energy balance model, as used by IPCC, is not new. It has been widely used to estimate magnitudes of gross energy exchange within the climate system. In these energy budget calculations the assumption of energy balance at the top of the atmosphere is a good approximation. Any actual imbalance at the top of the atmosphere is small when compared with the magnitudes of the major components of the global energy budget. Also, by averaging globally and over an annual cycle it is assumed that there is no need to consider seasonal accumulations of energy within the climate system nor the complex non-linear dynamics of the atmosphere and ocean circulations that transport energy around the globe. These assumptions are not valid in the context of climate change.

Overall, the energy balance model used by the IPCC reduces the complex climate system to a one-dimensional problem (see Figure 1). In this portrayal the climate system is made up of space, an atmospheric layer (the troposphere) and the earth's surface. Incoming solar radiation at the top of the atmosphere is balanced by the sum of the outgoing longwave radiation emitted by the earth's surface, cloud tops and greenhouse gases in the atmosphere, and of solar radiation reflected from the earth's surface, clouds and atmospheric aerosols.

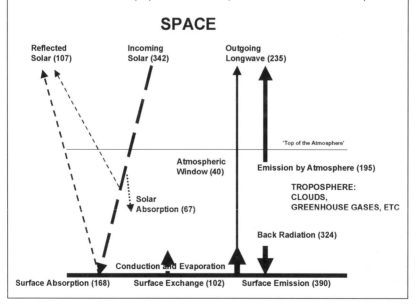

Figure 1:   The IPCC 'radiative forcing' model (Units – W/m²)

Put in its simplest form, the radiation forcing hypothesis used by the IPCC assumes that if the top of the atmosphere radiation balance is upset then the earth's surface temperature will adjust in order to restore the balance. Positive forcing (either an increase in solar input or a decrease in longwave emission) will lead to warming of the earth while negative forcing (a reduction in solar input or an increase in longwave emissions) will lead to cooling of the earth. The effect of increased concentrations of carbon dioxide in the atmosphere is to reduce the longwave radiation emissions to space. That is, energy will accumulate in the climate system causing a positive forcing; the reduced radiation to space will lead to global warming and the warmer temperatures will eventually increase the longwave radiation emission to space and restore radiation balance at the top of the atmosphere.

The radiation forcing hypothesis is simple, but inadequate. It portrays radiation as the central and only important process of the climate system. Those who ascribe to it have been seduced to forget elementary school geography; earth is a globe with seasonal patterns of solar heating that generate temperature differences between the tropics and the poles. The one-dimensional energy budget model is a prescription for flat earth physics whose application leads to erroneous conclusions.

It is essential to treat the earth as a sphere, recognising that almost nowhere is there radiation balance at the top of the atmosphere[6]. Most of the incoming solar radiation is intercepted in the tropics where, in magnitude, it far exceeds emission of longwave radiation to space. Over polar regions the emission of longwave radiation to space exceeds the intercepted solar radiation, especially during the darkness of winter. As a consequence, nowhere is the earth's surface temperature a function of local radiation characteristics alone. If it were not for the circulations of the atmosphere and the oceans the tropics would be hotter than they are and the polar regions would be much colder.

The net energy flow of the climate system is shown schematically in Figure 2. This shows important differences to the IPCC construct. Firstly, most of the solar radiation is intercepted by the earth over tropical latitudes, recognising that the earth is spherical. Secondly, the ocean surface layer is recognised as an important reservoir for heat storage. Thirdly, the circulations of the atmosphere and the oceans are important for transporting excess energy accumulating in the tropics to polar regions where there is a seasonal excess of radiation to space.

---

6   The exceptions are the zonal boundaries in each hemisphere that separates the tropics and their excess solar radiation from the middle and high latitudes where, except near midsummer, longwave radiation emissions exceed solar input.

The excess solar energy that is received over the tropics penetrates and is largely absorbed in the surface layers of the tropical oceans, forming an enormous energy reservoir. Energy is exchanged between the tropical ocean energy reservoir and the atmospheric boundary layer by conduction (sensible heat) and evaporation (latent energy). Deep tropical convection is necessary to distribute the latent and sensible heat from the equatorial boundary layer through the troposphere. The large overturning Hadley Cells of the tropics both converge heat and moisture of the boundary layer into the equatorial trough to supply energy for the deep convection and transport energy poleward in the high troposphere. Over the middle and high latitudes of each hemisphere energy is transported poleward by the atmospheric circulations associated with weather systems. Clouds and greenhouse gases in the troposphere emit longwave radiation to space. Over the poles the ice sheets act as an energy buffer; they absorb latent heat and

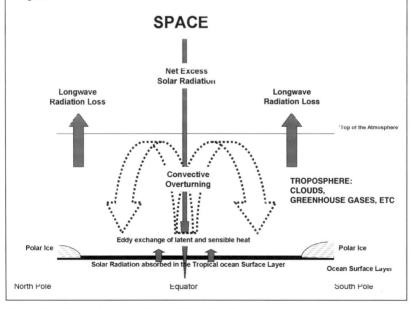

Figure 2:   The net energy flow of the climate system

melt during periods of transport excess and expand as accumulating snow releases latent heat during periods of transport deficit.

The magnitudes of the top of the atmosphere energy imbalances between the tropics and the poles ensure that there is a very large poleward transport of energy by the oceans and the atmosphere. However the oceans and atmosphere are fluids on a rotating earth and are subject to inertial oscillations and mutual interactions at the air-sea interface. A change of only a few percent in the poleward transport of energy by the atmosphere is sufficient to have a significant impact on the magnitude of the polar ice masses and surface temperatures. Natural variations of the ocean and atmospheric energy transports on interannual, decadal and longer timescales are, therefore, very important when considering the causes of surface temperature and polar ice volume changes. Over recent decades, monitoring of the impacts of El Niño events, particularly the major events of 1982–83 and 1997–98, has starkly illustrated the global impacts of the ENSO[7] phenomenon on climate, including enhanced export of energy from the tropics during El Niño events.

The IPCC has relied on computer models of the climate system for its projections of global temperature rise under the influence of increasing concentrations of atmospheric greenhouse gases. However these tools are currently inadequate for the task. The various models are fairly consistent in their ability to represent surface temperature from the tropics to polar regions, although increasing differences near the poles highlight deficiencies in representing the ice sheets. The various computer models do not as consistently represent surface air pressure as they do temperature, and again the differences are greatest near the poles. There are significant differences between the various models in their respective global precipitation distributions, including in the magnitude and patterns of equatorial rainfall. Precipitation is an outcome of atmospheric overturning and latent heat release and so the inconsistencies reflect differences in the internal circulation dynamics of the models. The representation of the ocean circulations, on average, is about half of that estimated for the climate system. This is a major defect of the computer models because it identifies that the transport of energy by the oceans is grossly underestimated. The surface net longwave radiation (one of the important components for anthropogenic greenhouse gas climate forcing) is also highly variable between models.

---

7   ENSO – El Niño Southern Oscillation, is the name given to a characteristic interannual variability of global atmospheric circulation patterns that is forced by changing sea surface temperature (SST) patterns of the central and eastern equatorial Pacific Ocean. An El Niño event is when SST over the eastern equatorial Pacific Ocean are warmer than normal.

The inadequacy of computer models in representing fundamental aspects of the climate system underscores the infancy of their development. Computer models that do not adequately represent the ocean and atmospheric circulations, quantities fundamental for accurately reflecting poleward energy transport, cannot be relied on to accurately project climate characteristics under changing scenarios of atmospheric greenhouse gases. The combination of a simple one-dimensional physical construct that is inadequate as a theoretical framework for anthropogenic greenhouse gas climate forcing and computer models that are in a rudimentary state of development has led the IPCC to false conclusions.

The climate system is very complex. The climate records that have accumulated in the archives of meteorological services clearly identify a warming trend at the surface around the globe over the 20th century. IPCC makes the claim that the recent warming is unusual, even unprecedented. The claim, supported by a dubious reconstruction of global temperatures over the past millennium and computer modelling, is the basis for the view that the warming, at least of the past 50 years, is due to anthropogenic greenhouse gases. Such a claim ignores the extent and speed of variability of past climate that can be interpreted from geophysical and biological evidence laid down in the oceans, lakes and ice sheets. It even contradicts cultural and archaeological records that reflect the rise of regional communities (even civilisations) during periods of favourable climate but their demise as conditions deteriorate.

Climate is naturally variable and it poses serious hazards for humankind. To focus on the chimera of anthropological greenhouse warming while ignoring real threats posed by natural variability of the climate system is self-delusion on a grand scale. In the following chapters we will look in detail at how climate has varied in the past; at the climate system and the natural processes that contribute to its variability; at the development of computer modelling of the climate system; and how the combination of a flawed theoretical framework and inadequate computer models have led IPCC to erroneous conclusions about future climate.

The Climate of Past Ages

Our knowledge of recent climate

Our personal perception of climate is largely developed through experience and interpretation of records compiled by our contemporaries and ancestors. People who grow up in the warmer temperate regions and the tropics are in awe at the first sight of snow, no matter what they have read or visualised from film and television. It is also a truism that what we perceive as normal climate for a locality is based on the weather we have experienced over recent years. This perception often occurs despite accounts of earlier catastrophes that had their origin in climatic extremes, such as a violent storm, flood or drought. Perhaps the exception is the markings seen around many riverside towns that point to levels achieved by past flood events. These are a constant reminder and it is very humbling to look upwards at the markings that are the remaining visual signs of inundation, destruction and even deaths from previous centuries.

One of the strengths of humankind has been the ability to survive, adapt and prosper across a wide range of climatic regimes. Historically, communities have also shown a capacity to withstand persisting climatic fluctuations. This is not to say that extended drought or periods of warmth or cold do not have an impact on social systems. They do, but resilient communities survive and recover when climatic conditions return to what is more typical of the long-term patterns that the social system is accustomed to. However, as we will see, there have been times when prosperous civilisations have fallen, apparently because the regional climate change was so severe and prolonged that the social systems based on food production and trade could not be sustained.

We recognise seasons and acknowledge that there is year-to-year variability in our climate. It is for this reason that recording of changing annual events has been a feature of community life over several thousand years. Early records of the annual flow of the Nile River more than two thousand years ago and irrigation activities in China more than one thousand years ago survive and give us insights into how climate has been in the past. Beginning about a thousand years ago, there is an array of phenological data relating to community life in Europe that provides information about how local and regional climate might have changed with time[8]. These records tell us a lot about the close linkage between the annual climatic regime and the social infrastructures that developed.

It was not until the invention of practical meteorological instruments, including the thermometer and barometer during the 17th century, and the acceptance of standard measuring scales in the early 18th century, that quantitative comparisons of the climate between different locations could begin to be made. Observations from different localities showed the extent of variation from place to place and systematic records from a particular locality spanning many years were used to characterise the seasons, the annual climate cycle and its variability from year to year.

By the beginning of the 19th century systematic meteorological observations, based on practical instruments to measure primary meteorological elements, were being made and recorded at a number of scientific observatories across Europe and settled parts of the Americas. The meteorological observations, including temperature, pressure and rainfall, were often part of a broader scientific program in conjunction with other geophysical observations, including astronomy and measurements of geomagnetism. The observations contributed to broader scientific investigations of the natural world that were taking place.

In the early years of the 19th century there were relatively few observatories. However scientists began exchanging data to map concurrent meteorological observations from different locations and analyse the location and intensity of weather systems. These early studies showed, in a rudimentary fashion, the scale of weather systems and the direction and speed with which they moved. As a consequence governments began to organise programs for the systematic observation of weather and the collection and analysis of the data to provide warning of dangerous storms. Under government direction the number of meteorological observing sites increased rapidly as local officials were co-opted to make regular observations.

By the middle of the 19th century there was a developing recognition of benefits from systematically monitoring the local

---

8   While a good portion of the phenological data came from monitoring local agricultural events some records seem to be more esoteric, such as relating to the first sighting of migratory birds, the flowering of particular plants, etc.

climate[9]. Studies were being made to identify the optimal climate for different agricultural and horticultural plants and how production varied under different seasonal conditions. Statistics, as a branch of mathematics, owes much of its early development to studies involving meteorology and agriculture.

Mariners, who were constantly threatened by storm and tempest as they sailed the oceans, were particularly aware of the necessity for maintaining watch of the changing meteorological conditions. A body of sea lore had developed that provided characteristic early warning of storms. Rapidly changing atmospheric pressure and changes in wind direction were noted as harbingers of danger. It was Lt Matthew Maury of the US Navy who pressed the need for consistent instruments and recording methods to ensure that data from different locations could be compared accurately. Maury played a leading role in organising the first International Meteorological Congress of 1853, held in Brussels, to establish standards for making meteorological observations at sea and recording the data in Ships' Logs.

The collection and analysis of the early meteorological observations made by ships and from the land stations provided the basis for our earliest understanding of the climate patterns of the world. Vladimir Koppen, Director of the Naval Observatory of Hamburg, spent more than a quarter of a century gathering and analysing world-wide meteorological data and descriptions of local vegetation. Koppen's first Climatological Atlas was published at the turn of the 20th century and identified a range of vegetation types with characteristic temperature and rainfall patterns.

What this brief history of meteorological observing demonstrates is that our systematic knowledge of climate covers little more than a hundred years. But even our 'instrumental record' of climate can be misleading unless analysed with care. The series of records from before the 20th century often were from a sequence of different instruments reflecting advancement in technologies, and each instrument had its own response characteristics. For thermometers, alcohol in a sealed thin glass column was initially the preferred expansion fluid and responded rapidly to changing temperature. Mercury-in-glass thermometers then became widely used because of their ability to

---

9   During a period of residence in Switzerland the author was fascinated by the prominent location of meteorological instruments of considerable age that featured in most small towns. Always a large thermometer, mostly a barometer and often an hygrometer set into a stone column or attached to a wall at a central location. These instruments have their origin in the middle 19th century and reflect the close linkage of agriculture to climate. Touring through the mountain valleys was slowed by the need to stop and find the local observing site when passing through a village.

measure across the wide range of temperatures that are experienced from the cold polar regions to the hot deserts. However, because mercury is denser and has a higher thermal capacity than alcohol, the mercury-in-glass thermometers respond more slowly to changing temperatures. Also, it was not until the late 19th century that standard meteorological shelters for housing the outdoors instruments were adopted worldwide.

Basic meteorological instruments and their shelters remained unchanged for most of the 20th century. The large white wooden instrument shelters, and their slatted sides for airflow, have been common sights. What have not been so obvious are the slow changes that have taken place around most observing sites as villages grew into towns and urban development has changed environmental characteristics. Of particular significance has been the replacement of trees that draw on deep soil moisture to maintain evapotranspiration and provide cooling during hot weather. The substituted concrete and asphalt of buildings and paving do not evaporate soil moisture and they readily absorb solar radiation. Therefore, a slow warming of the urban environments is being registered and can be detected in the long meteorological records.

Another factor related to observing practice that is adding to our uncertainty of how the climate might be changing, is the so-called modernisation of meteorological instruments. Routine observing of the instruments and recording the data several times a day, seven days a week requires dedication. Many National Meteorological Services began installing automated recording instruments during the last two decades of the 20th century. These instruments have been so successful that their use has proliferated and they are becoming common worldwide. The automated instruments again have different response characteristics to those that they replaced. For temperature, the physical characteristic used is the changing electrical conductance of an exposed element rather than the expansion of mercury. The newer instruments are much more responsive to changing temperature than the mercury-in-glass thermometers.

It is not only the land stations for which the changing instruments and observing environment have affected the meteorological quantity being observed. Ships' observations are compromised by the location of the instrument shelter relative to sea level as the size of vessels increased. Merchant ships are the primary source of meteorological observations from the oceans and the modern bulk carriers and container ships tower over the cargo ships from the late 19th and early 20th century. Air temperature decreases with elevation so that the gradual increase in the observing height above sea level has introduced a small bias to marine observations.

Merchant vessels are also the source of our estimate of sea surface temperature, one of the most extensive observations from the oceans and important for understanding the energy exchanges between the oceans and the atmosphere. Very early estimates of sea surface temperature were obtained by collecting a bucket of water from over the ship's side and measuring its temperature. This method was prone to inaccuracy because the water temperature adjusted to the air temperature and contributed to error if there was a delay in measurement when the bucket of water was lifted to the deck. A more practical method that has been used for more than a century is to note the temperature of the intake water for cooling the ship's engines. However, the depth below the surface of the intake pipe is not standard and also varies with each ship depending on whether it is empty or fully laden. This inconsistency of the measuring depth creates uncertainty because ocean temperature often changes significantly with depth close to the surface.

Surface temperature is the most widely used statistic for describing climate change because of the length of systematic instrumental records and the spatial density of observing sites that are available. The linkages with agricultural production and weather forecasting have ensured a steady increase in the number and spatial density of land-based observing sites, at least until the 1970s. The collection and analysis of meteorological data from ships' logs has provided data from those parts of the oceans covered by the major shipping routes. Since the early 1960s, ships data have been supplemented by data collected from drifting meteorological buoys scattered across the world oceans.

For the past two decades the ships observations of sea surface temperature have also been supplemented by observations based on data from satellite-borne instruments. The satellite-based observations provide complete coverage of the world oceans. The satellite-borne instruments record the intensity of radiation in a narrow wavelength emitted from the sea surface. The satellite instruments are calibrated against ship and drifting buoy observations and the data used to extend sea surface temperature patterns across the globe.

Overall, however, what is referred to as 'surface temperature' is generally a combination of entities. Over land it is the near-surface air temperature measured at meteorological observing sites. These data have a marked diurnal range and the daily surface temperature is often the average of the afternoon maximum and early morning minimum temperatures. Over the ocean the diurnal range is more limited but instruments are more likely to be affected by solar radiation. Therefore the surface temperature over the ocean is often represented by the nighttime air temperature (from ships or buoys) or sea surface temperature.

The meteorological records gathered from land and the oceans over the past century require careful analysis if they are to be used for identifying real changes in climate and not the artefacts of the measuring instruments or their environment. Nevertheless, the records do provide a wealth of data from which a range of conclusions has been drawn when the non-climatological factors described above are extracted. Several analyses of the global temperature records have been performed and, although different researchers have arrived at slightly different values, a common warming trend during the 20th century is evident. One estimate, compiled by the United Kingdom Meteorological Office and adopted by the IPCC, is shown in Figure 3. The graph shows the global average departure of the surface temperature at each location from its average for a standard period 1961–1990. The solid line is the moving 10-year average and gives a smooth trend with time.

The global surface temperature over the past 140 years has changed. There are significant year-to-year fluctuations throughout the record but superimposed on the short-term variability is an overall warming trend. The warming is not uniform but confined to the periods 1910–1942 and 1978–2000. Prior to 1910 there was no discernible long-term trend and the period 1943–1977 was characterised by a slight cooling.

The emphasis on global surface temperature in discussions on climate change reflects the difficulty in characterising the earth's

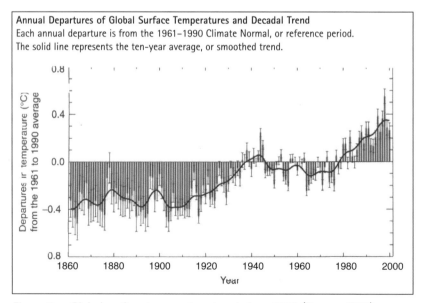

Figure 3:   Global surface temperature trend since 1860 (Source: IPCC)

climate system with a simple unambiguous index. Surface temperature, despite its limitations, is the most spatially representative meteorological variable with a relatively long record. However the temperature of our environment is only one of many key meteorological factors that are important for life on earth as we know it. Plant growth responds to warmth but is also dependent on sunlight, moisture, the availability of nutrients (often carried with soil moisture) and carbon dioxide. Only a few of these important quantities are measured routinely and then from limited land sites.

It is only for the past few decades that satellite-borne instruments have been available to routinely measure, with global coverage, a range of climatologically important quantities. These data are assisting the management of many land and ocean based climate sensitive industries. They are also providing important information that is leading to a better understanding of how the climate system functions. In particular, various sensors monitor the global distribution of net radiation at the top of the atmosphere while other sensors contribute to the assessment of heat, moisture and momentum within the atmosphere.

It will be decades before the new satellite data will provide unequivocal assessments that differentiate between apparent long term trends in the overall climate system and multi-decadal variability. Until that differentiation is made there will be continuing speculation over whether the recent apparent increase in global surface temperature represents an unusual trend forced by human activity or whether it is part of the natural variability of the climate system. If there is significant natural variability of the climate system, either as a response to internal processes or as a consequence of solar forcing, then we might expect evidence of this in the palaeontological evidence retained in the remnant life forms of earlier times.

Reconstructing past climate

Knowledge about past climates is not entirely lost. Although there are no direct measurements of important climatological quantities there is evidence of past climatic regimes contained in the sediments laid down on ocean and lake floors, in the accumulated annual snowfall preserved in glaciers and in the growth patterns preserved in wood and coral. New technologies are leading to the identification, timing and magnitude of past climatic shifts. Over recent years, more palaeoclimate data with improved time resolution have been gathered to give a better picture of how earth's climate has varied in the past.

Drilling and recovery of deep cores from polar ice sheets, mountain glaciers, the sea bed and lake floors provide relatively long continuous chronologies of how factors associated with formation of the ice and sediment layers have varied. Where the ice sheets and sediments accumulate in annual layers the timing between events, and often the actual age, can be determined with precision. Other methods, such as those based on the decay of radioactive isotopes, are available to date material where annual layering is not discernible. In many cases the time resolution may be no better than decadal or centennial, depending on the accumulation rate of the material being sampled and its overall age.

The physics, chemistry and biological origins of material accumulated at successive depths of ice or sediment provide insights into the prevailing climatic and environmental conditions as the materials were being deposited. The composition of air bubbles trapped in the ice, and the composition of dust, silt and biological materials, including the relative populations of skeletal species, all contribute to our knowledge of past climates. The recovered cores each give a local record back in time that can extend over thousands to millions of years.

Continuous cores retrieved from deep ocean drilling and from the deep ice over Greenland and Antarctica are the basis for the longest relatively detailed climate records. The ice cores have provided information on the timing and relative intensity of previous glacial and interglacial epochs as well as a number of changing atmospheric characteristics, such as carbon dioxide and methane concentrations, during these times. The cores from the ocean floor often contain material indicative of regional and more local conditions, including rate of accumulation of blowing dust from an adjacent continent (relative aridity), relative abundance of different plankton species (ocean surface temperature) and ice rafting debris (changing ocean currents). The changing isotope ratios of various chemical elements in the trapped materials also contribute to our understanding of past climates.

Annual tree rings, and their variation in width and density that correspond with the changing temperature and rainfall from year to year, are a primary source of information over recent centuries. The annual growth layers of coral also vary with environmental conditions. Care must be taken with the interpretation of these data because the annual variation may result from one or more of several factors. For example, in very cold climates the tree growth only occurs over the warmer summer months and any variation does not represent changes in winter conditions. Similarly, in semi-arid climates the temperature signal may be confounded by the varying rainfall signal.

There are many sources of uncertainty in the reconstruction of a climatic history from proxy data, particularly for the oldest data. Interpretive methods are often necessary in the dating of individual events. In records where there is poor time resolution the characteristics can only be attributed to a period of years, or even decades. Where there is a very strong annual signal, such as with tree ring and coral growth layers, precise timing intervals can be given between particular events. The accumulation of ocean sediments is often not in identifiable annual layers and the rate of accumulation may vary with environmental conditions. When reconstructing the time history of a climate proxy, account must be taken of the changing time resolution and signal fidelity of the proxy being examined. This is also important in climate reconstruction that uses different proxies over the interval of interest.

There is a wealth of climate information in the palaeoclimate records but, even with currently available technology, the reconstructions lack the sensitivity and precision of modern instruments. As we go back further in time the resolution deteriorates as the signal is smoothed over ever expanding time intervals. Thus, we can resolve annual change in the instrument record and some of the millennium scale reconstructions of temperature. The palaeoclimate data are not as sensitive as instruments and the magnitudes of extremes are underestimated. The farther we go back in time the less detail that can be resolved from the record. It is only possible initially to resolve climate shifts that occurred over decadal timescales, then centennial timescales, and so to characteristics that vary significantly on ever lengthening time intervals.

In the following sections we will review the general characteristics of the varying climate as revealed by palaeontologic evidence, including the glacial epochs, the Holocene[10] since the last major glaciation of the earth, and particularly the last millennium with its historical and cultural records. In doing so, we should keep in mind that the absence of detail for the earlier period does not mean that short-period climate fluctuations did not happen – it is just that we do not have information about them.

There is no single source of information for the earth's climate before the instrumental period beginning in the mid-19th century. Similarly, reconstruction of past climate was not a steady development through the accumulation of more and more data. Rather, our understanding of past climate has progressed in steps as new technologies and data have advanced and often forced the discarding of previously accepted chronologies.

---

10 'Holocene' refers to the current interglacial period that started about 10,000 years ago.

Ice Ages and lesser climate fluctuations

The cornerstone of reference for modern climatology is the recurring glacial epochs that have been a feature for the past several million years. We are currently experiencing an interglacial epoch and the earth is apparently near the warmest it has been during the past million years. Currently, the extent of mountain glaciers and polar ice sheets are significantly reduced and sea level is about 130 metres above the elevation during the last glacial maximum about 15,000 years ago. However there is evidence that the climate has been warmer and sea level from 3 to 20 metres higher than it currently is during several previous interglacial epochs.

Past glacial periods left their evidence as scoured valley walls and remnant terminal moraines. For much of the 19th and early 20th centuries evidence was gathered during field observations at many locations around the world. The data were assessed and estimates were made of the geographic extent of Northern Hemisphere glaciation. There were also estimates made of the depth of continental ice sheets that formed over North America and Europe. What were missing from these early observations and assessments were good estimates for the timing of the major periods of glaciation.

A range of new technologies became available in the second half of the 20th century that provided the basis for better timing of changes in climatic indicators. The drilling and recovery of ice cores from Greenland provided a chronology going back more than 100,000 years. Later drilling of Greenland and Antarctica extended the record to about 450,000 years. It is the polar ice cores that have provided evidence of the regularity of glacial epochs and the temperature changes that have occurred with them. Figure 4 is a record of the

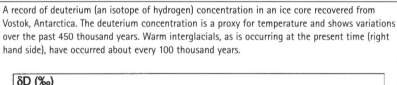

A record of deuterium (an isotope of hydrogen) concentration in an ice core recovered from Vostok, Antarctica. The deuterium concentration is a proxy for temperature and shows variations over the past 450 thousand years. Warm interglacials, as is occurring at the present time (right hand side), have occurred about every 100 thousand years.

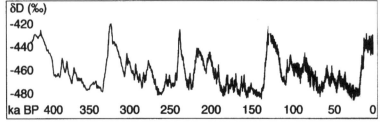

Figure 4:   Antarctic temperature proxy recording glacial and interglacial periods about every 100 thousand years

deuterium ratio (deuterium is an isotope of hydrogen) in the ice core from Vostok, Antarctica and maps the changing temperature over the past 450,000 years. There is a recurring pattern of cooling and expansion of polar ice over a period of about 80,000 years followed by a warming and contraction lasting about 20,000 years. The approximately 100,000 year pattern is evident from ice cores covering the past 450,000 years and sediment cores from the ocean floor suggest the pattern has repeated over several million years.

The sheer scale of the glacial epochs means that they have been of great interest to scientists. As we shall see presently, trying to explain the occurrence of glacial periods has been the focus of many climate theories. However, within the broad envelope of the glacial and inter-glacial epochs there is much more variability. This shorter period variability is of interest because, in more recent times, it comes down to timeframes relevant to modern civilisations and how they have been affected.

Figure 5 shows data analysed from a central Greenland ice core for the period since the last glacial maximum, nearly 15,000 years ago, and identifies intervals of rapid change, both during the warming phase and the relatively stable climate period from the last 8,500 years of the Holocene. For example, there is evidence that the Greenland Summit (site of ice drilling for the Greenland Ice Sheet Project – GISP2) warmed about 9°C over several decades about 14,500 years ago[11]. Ice cores from the Antarctic ice sheet generally reveal a similar warming phase but it does not appear that the Antarctic Cold Reversal is fully in concert with the Greenland warming[12]. An exception is a core from Taylor Dome, in close proximity to the Ross Sea, where major fluctuations are synchronous with Greenland events[13].

The Greenland ice core also shows that later in the warming phase, about 12,500 years ago, temperatures plummeted to near ice age conditions, and that the cold lasted nearly 1,000 years (a period known as the Younger Dryas). The Younger Dryas event has been linked to the melting of the North American ice sheet, specifically the sudden diversion of freshwater ice-melt from the Mississippi drainage basin to

11 Severinghaus, J.P. and E.J. Brook, 1999. Abrupt climate change at the end of the last glacial period inferred from trapped air in polar ice. Science, vol 286 pp930–934.

12 Morgan, V., M. Delmotte, T. Van Ommen, J Jouzel, J. Chappellaz, S. Woon, V. Masson-Delmotte and D. Raynaud, 2002. Relative timing of deglaciatial climate events in Antarctica and Greenland. Science, vol 297 pp1862–1864.

13 Steig, E.J, E.J. Brook, J.W.C. White, C.M. Sucher, M.L. Bender, S.J. Lehman, D.L. Morse, E.D. Waddington and G.D. Clow, 1998. Synchronous climate changes in Antarctica and the North Atlantic. Science, vol 282 pp92–95.

14 Broecker, W.S., 2003. Does the trigger for abrupt climate change reside in the ocean or in the atmosphere? Science, vol 300 pp1519–1522.

the St Lawrence outflow into the North Atlantic[14]. The sudden input of freshwater into the North Atlantic Ocean is hypothesised to have interrupted the thermohaline circulation[15] of the ocean to such an extent that there was strong feedback to the climate system, at least across the Atlantic basin. However the processes which bring about the feedback and its long and severe impact are not yet fully explained[16].

Two deep-sea cores from opposite sides of the North Atlantic, and more than 1,000km apart, reveal that near synchronous abrupt climate

A record of Temperature (upper line) and snow accumulation rate (lower line) for the past 17 thousand years estimated from an ice core recovered from Greenland. Recovery from the last glacial maximum commenced about 15 thousand years ago before the return to glacial conditions (the Younger-Dryas). Warming again commenced about 12 thousand years ago and the interglacial warmth has persisted for the past 8 thousand years (the Holocene). Temperature has fluctuated through the Holocene.

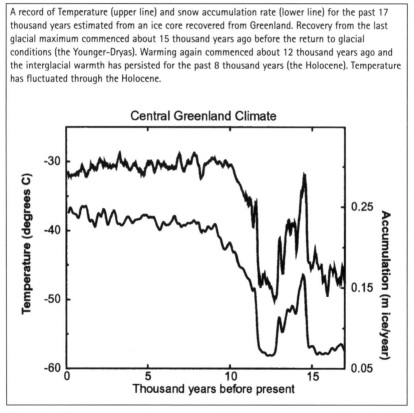

Figure 5: Temperature and snow accumulation rate over central Greenland as reconstructed from an ice core

15 The thermohaline circulation is the slow overturning of the ocean brought about by cooling and sinking of very saline and near-freezing waters over the polar oceans. The process is explained in more detail later in the discussion on ocean circulations.

16 Stocker, T.F., 2002. North-South connections. Science, vol 297 pp1814–1815.

shifts punctuated the Holocene, a period of relative warmth not too dissimilar to the present climate[17]. The cores reveal that ice-rafted debris events recur on millennium scales with peaks at about 1,400, 2,800, 4,200, 5,900, 8,100, 9,400, 10,300 and 11,100 years ago. The ice-rafted debris occurs because ice sheets moving over land scour the underlying surface and carry embedded material. Rafted ice (icebergs) is calved from the coastal glaciers and drift through warmer waters. As the ice-rafts melt the material scoured from the land and embedded in the ice is released and settles to the ocean floor. The physical and chemical composition of the ice-rafted debris can point to the origins of the calving. The increase in ice-rafted debris during these millennial-scale events is consistent with a significant change in climate and ocean circulation over the North Atlantic Ocean.

During periods of relative warmth, similar to those currently prevailing, icebergs entering the sub-polar North Atlantic are calved in southern and western Greenland and move in the southwesterly gyre towards Newfoundland. During the colder events, however, the icebergs have their origins in the Greenland-Iceland Seas and further north around the Arctic Ocean. They are transported further south and southeastward because of a shift in the currents of the cold polar and sub-polar surface waters. The decrease in sea surface temperature over the North Atlantic during these events is estimated to be only about 2ºC cooler than the present. Although not identified in the ice-cores, the documented increase in observed Icelandic coastal sea ice around about 1800 AD is consistent with these earlier events of relative coldness[18].

Corroborating evidence for millennial scale climate variability over the North Atlantic Ocean region is found in variations of the speed of deep ocean currents south of Iceland[19]. The grain size in silt laid down during the Holocene has been analysed from sediment cores drilled from the ocean floor. Varying grain size is used as a proxy for bottom-water current and a measure of the strength of the thermohaline circulation. Flow changes correspond with the documented Little Ice Age and Medieval Warm Period of northern Europe and extend over the entire Holocene epoch, with a quasi-periodicity of about 1,500 years. The sediment indicators point to faster bottom-water flow from the Arctic during warmer intervals over most of the Holocene. There is, however, a reversal of this relationship

17 Bond, G., W. Showers, M. Cheseby, R. Lotti, P. Almasi, P. deMenocal, P. Priore, H. Cullen, I. Hajdas, and G. Bonani, 1997. A pervasive millennium-scale cycle in North Atlantic Holocene and glacial climates. Science, vol 278 pp1257–1266.
18 Grove, J.M., 1988. The Little Ice Age. Methuen, London.
19 Bianchi, G.G. and I.N. McCave, 1999. Holocene periodicity in North Atlantic climate and deep-ocean flow south of Iceland. Nature, vol 397 pp315–317.

earlier in the period. It is speculated that the reversal is a consequence of remnant land-based glacial ice at that time contributing fresh water to the sub-polar North Atlantic Ocean and interrupting the thermohaline production of deep water.

The sediments laid down on a lake floor in the high Andes Mountains of Ecuador have also been used to identify fluctuations of climate through the Holocene[20]. During modern times it has been found that rainfall over the high Andes Mountains increases markedly during El Niño events[21] over the coastal and mountain regions of Ecuador and northern Peru. The high rainfall rates during these events increase soil erosion across the drainage basins and change the physical and biological characteristics of silt deposition in high lakes. A continuous sediment core covering the recent 12,000 years has been interpreted as a record of the occurrence of major El Niño events and suggests an increasing rate of occurrence from about 7,000 years ago, reaching a maximum about 1,000 years ago. Superimposed within this envelope is millennial scale variability with a cycle of about 2,000 years early in the Holocene that decreases to about 1,500 years later in the period.

Analysis of the physical and chemical properties of ice cores drilled in the remnant ice sheets of Mt Kilimanjaro, in equatorial Africa, identify a complex climatic history over the recent 11,700 years of the record, including periods of abrupt climate change about 8,300, 5,200 and 4,000 years ago[22]. The ice sheets receded significantly over the 20th century and are likely to disappear within decades if the current rate of contraction persists. Enrichment of isotopic oxygen and reduced concentrations of aerosol species indicate warmer and wetter conditions were dominant from about 11,000 to 4,000 years ago and are consistent with a prolonged humid period identified in records from elsewhere in Africa.

High concentrations of sodium and fluorine ions in the Kilimanjaro ice core about 8,300 ago are consistent with a large and rapid drop of nearby lake levels (enriched with these ions). It is

20 Moy, C.M., G.O. Seltzer, D.T. Rodbell and D.M. Anderson, 2002. Variability of El Nino/Southern Oscillation activity on millennial timescales during the Holocene Epoch. Nature, vol 420 pp162–165.

21 In most years, rainfall over the coastal plains and nearby mountains of Ecuador and northern Peru is very low and the landscape is arid. During El Niño events the nearby coastal waters are anomalously warm and provide moisture to sustain heavy tropical rainfall over the coastal lowlands and nearby mountains.

22 Thompson, L.E., E. Mosley-Thompson, M.E. Davis, K.A. Henderson, H.H. Brecher, V.S. Zagorodnov, T.A. Mashiotta, P-N. Lin, V.N. Mikhalenko, D.R. Hardy and J. Beer, 2002. Kilimanjaro ice core records: evidence of Holocene climate change in tropical Africa. Science, vol 298 pp589–593.

suggested that wind erosion increased the atmospheric loading of the particular aerosol species, and hence the deposition rates on the Kilimanjaro ice fields. The event is coincident with a drop in methane concentrations observed in the Greenland ice cores that has also been attributed to tropical drying, particularly in Africa.

Lower isotopic oxygen concentrations in the Kilimanjaro ice cores 6,500 years ago are interpreted as a cooler humid period when conditions were wetter than today but drier than the early Holocene. A sudden drop in oxygen isotope concentration about 5,200 years ago is coincident with a century scale period of cooler and drier conditions. Documentary evidence[23] suggests that drought was widespread over Africa and the Middle East about the time indicated in the ice core. Atmospheric methane concentrations, as recorded in Greenland ice cores, reached their lowest levels during the period, further supporting the view of prevailing dry conditions. Warmer and more humid conditions then persisted until about 4,000 years ago.

The Kilimanjaro ice cores indicate that a warm and very dry event occurred about 4,000 years ago. Parts of the currently existing ice sheets were not present at that time and the cores that did record the event featured a distinct visible dust layer. High concentrations of other chemical species in the layer suggest a hiatus in ice-mass accumulation. Evidence is advanced[24], including an enormous dust event recorded concurrently in the ice cores of the Andes Mountains of northern Peru, that suggests an extensive and extremely severe 300-year drought occurred.

Further evidence of a dramatic widespread shift in the climate of Africa about 5,500 years ago comes from an ocean sediment core recovered from the Atlantic Ocean off northwest Africa (at latitude 20°N)[25]. The marked increase in blown Saharan mineral aerosol dust indicates that North Africa became more arid with extensive desertification after that time. There was an almost concurrent rise in sea surface temperatures that has been attributed to a slackening of the tropical easterly Trade Winds and a reduction in upwelling of deep colder water off the West African coast.

The solar insolation[26] of summer has been steadily reducing over the northern hemisphere from about 10,000 years ago as a

---

23 H. Weiss, quoted in Thompson et al, 2002. Ibid.
24 Thompson et al, 2002. Ibid.
25 deMenocal, P., J.Ortiz, T. Guilderson and M. Sarnthein, 2000. Coherent high- and low-latitude climate variability during the Holocene warm period. Science, vol 288 pp 2198–2202.
26 Solar insolation is the accumulated daily solar energy received at a location. Although the intensity of solar radiation is much less, the summertime solar insolation is slightly more over the polar regions than over the tropics. This is because of the much longer daylength over polar regions in summer.

consequence of the precession of the earth's axis. It is not clear why the climate of the subtropical regions of North Africa should suddenly flip from a wetter to a drier state about midway in the cycle.

Analysis of the Atlantic sediment core from off the northwest African coast also identifies millennial scale variability in sea surface temperatures through the Holocene period[27]. The variability is synchronous with the ice-rafting debris events of the North Atlantic that have been referred to earlier. The question remains as to whether the cooling events should be attributed to changing strength of the Trade Winds, and hence varying upwelling of colder sub-surface water, or changes in the strength of the eastern Atlantic boundary current (the Canary Current) that transports colder waters southward[28].

This brief description represents only a few of the many results from palaeoclimatic studies that are available. It demonstrates how localities have been subjected to ongoing variability and sudden shifts in climate during the Holocene period.

The human dimensions of climate change

The Holocene has been referred to as one of climate optimum but this is only in comparison to the earlier glacial epochs. However, any impression of climate stability or uniformity in which civilisations flourished without hindrance should be dissuaded. The climate variability during the Holocene that we have previously noted has had severe impact on the biosphere and human civilisations that were developing during the period.

Most archaeological evidence of early civilisations and their development comes from records covering the Holocene period as the climate recovered from the last glacial maximum. For the past 8,500 years, global climate has varied around conditions not too dissimilar to the present, but sufficiently different to have severe impact at times in some locations. Sudden climatic shifts, such as from humid to arid conditions over Africa about 5,500 years ago, would have had dramatic consequences for local communities, both nomadic peoples and those engaged in early agriculture. It is important for modern communities that as they plan for the future they are aware of natural changes that have occurred in the past and are likely to occur in the future.

There is abundant evidence from various parts of the world that early civilisations and the development of modern civilisations were affected significantly by changing local climates. We have already noted the rapid drying over the now arid regions of North Africa and the Middle East from what was more plentiful rainfall and open

27 deMenocal et al, 2000. Ibid
28 deMenocal et al, 2000. Ibid

grassland and woodland landscapes. The extremely dry, century-long period of blowing dust over North Africa about 5,000 years ago would have harshly treated hunter-gatherer populations and any but the hardiest communities relying on primitive agriculture.

The Atacama Desert region of northern Chile[29] provides one example of how regional populations initially flourished but then vacated landscapes. In modern times the rainfall ranges from 20 mm per year at lower elevations to 200 mm per year over the higher altiplano. Vegetation is extremely sparse. There is widespread evidence of settlements over the area from 13,000 years until about 9,500 years ago, mainly at higher elevations in association with late-glacial lakes on the altiplano but also in association with wetlands at lower elevations. Cessation of occupation of many sites between 9,500 and 4,500 years ago is linked to drying of the lakes under conditions of reduced rainfall. The dryness was widespread through the region with low lake levels across the south central Andes and Titicama basin. There were also low accumulation rates and elevated levels of chlorine ions in Andean ice cores covering this period. With the return of wetter conditions about 4,500 years ago the settlements were again occupied.

Over a period of more than 20 years from the mid-1960s to the mid-1980s, the English climatologist Hubert Lamb published extensive material relating to climate and civilisations, mainly relating to the Eurasian continents[30]. He has described development and adaptation as the middle and high latitudes emerged from the icy depths of the glacial maximum, ice sheets receded, sea level rose and vegetation changed. Over the subtropics temperature was not so decisive for habitation and it was rainfall, or lack of it, that regulated habitation patterns.

Reconstructions of the atmospheric circulation prevailing during stages of the Holocene suggest that between 9,000 and 5,000 years ago the subtropical anticyclones were further north reaching around latitude 40°N through the middle of the period. Such a poleward excursion would have allowed summer monsoons to penetrate further northward, bringing enhanced summer rainfall and supporting more abundant vegetation and wildlife. Over the Sahara, Lake Chad was more extensive than today. Also, the water levels of Lake Rudolf and other lakes of East Africa were some tens of metres higher and the landscape has been described as Savanna-like. Areas of the Sahara

29 Núñez, L., M. Grosjean and I. Cartajena, 2002. Human occupations and climate change in the Puna de Atacama, Chile. Science, vol 298, pp821–824.
30 See, for example: Lamb, H.H., 1995, Climate, History and the Modern World. Routledge, Second Edition pp431, from which material for this section was drawn.

that today rarely experience rainfall are estimated to have received annual rainfall between 200 and 400 mm.

Over parts of North Africa and the Middle East there is evidence that the wetter regime was interrupted by nearly 1,000 years of extremely dry conditions starting about 8,000 years ago. Archaeology from the Palestine region suggests that expansion of human settlements into the driest regions seems to have had two highpoints – about 8,000 years ago and 6–5,000 years ago.

The period around 5,000 years ago was one of rapid climate change across the southern Eurasian continent to which populations had to adapt. A great swathe from Arabia, through Afghanistan, Rajastan and Sinkiang, to the Gobi was becoming more desert like. As pastures and stocks of wild game failed populations were concentrated into the valleys and well watered lowlands to take advantage of more reliable water supplies. The forced settlement is offered as one of the reasons for the development of social cohesion and ordered communities, as well as the introduction of irrigated agriculture along the vast waterways of the main river systems.

It has been estimated that sea level was at its highest about 4,000 years ago. Coastal inundation would have altered present shorelines and accounts for coastal dunes that are well inland today. It was about this time that a freshwater canal was constructed to link the Nile to the Red Sea at Suez, taking advantage of a shortened land barrier.

There has been a trend of general cooling since the temperature peaked 5,000 years ago but the cooling has not been regular and not altogether synchronous worldwide. A number of abrupt changes that lasted from a few years to a few centuries occurred in some localities. Tree lines reached their maximum poleward latitudes or mountain elevations and have not returned to levels prevailing at the time of the optimum. The general drying of the North African and southern Eurasian subtropics after about 5,000 years ago does seem to have occurred synchronously with other significant changes around the world. For example, there is also evidence of drying over the Australian continent commencing about this time.

Over the southwestern United States (New Mexico and Arizona) agriculture had been expanding northwards in a wetter period between 6,500 and 6,000 years ago. However there is no evidence of agriculture over the region for the next 2,000 years, indicative of the onset and persistence of a substantially drier period and the abandoning of farmlands.

Egyptian records, inscribed on stone tablets, identify very dry periods and great famines lasting many decades that occurred about 4,150, 4,000 and 3,800 years ago. It is recorded that these droughts and famine were associated with very low flows of the Nile and a prevalence of southerly winds. These were also times when Egypt was invaded by peoples from the east, suggesting widespread drought may have been unsettling neighbouring communities to the point of open warfare for limited food resources.

About 3,500 years ago another significant change-point in climatic history took place as renewed cooling set in and mountain glaciers advanced worldwide. After this time there were significant migrations of peoples, both southward over southern Europe and eastward from the Hungarian Plain into Asia Minor. Distinctive settlements, which had been built on piles in the edges of lakes during the preceding warmer time, were abandoned, possibly by catastrophic flooding episodes. In the subsequent warmer and drier period of about 3,000 years ago new lake settlements were established. Peoples from the west and northeast again threatened Egypt about 3,200 years ago as drought prevailed across the eastern Mediterranean. There are historical accounts of famine and pestilence affecting the island of Crete to the extent that it was almost uninhabitable.

A period of colder and wetter conditions set in over Europe between 2,800 and 2,700 years ago. Norwegian glaciers advanced, reaching positions almost as far forward as during the worst periods of recent centuries. Nearly 2,500 years ago, Lake Constance rose rapidly by about 10 metres and all the lake settlements were abandoned. The alpine populations seem to have fallen to a minimum and were largely confined to the warmer valleys. High level mining that had been flourishing was brought to an end because of the advance of glaciers.

Over China the warmth of the post-glacial period came to an end rather abruptly between 3,100 and 2,800 years ago. Bamboo retreated from northern parts, the dates of rice and fruit harvests became later and freezing of the Han River in Central China was recorded. These are conditions similar to those currently prevailing. A warmer period lasting nearly six centuries prevailed before another onset of cold conditions about 2,200 years ago.

Archaeological evidence from the Valley of Mexico suggests that there were three periods of maximum human settlement before the arrival of Europeans. These were the periods from 2,500 to 2,100 years ago, about 800 AD and the three or four centuries of the Aztec period up to 1500 AD. These periods coincide with the main periods of colder climate over Europe and much of North America. However it was probably additional moisture that was beneficial for Mexico. Settlements were able to spread from the lakeshore plain to the foothills, where the soil is thin and the climate today is too dry for most kinds of agriculture.

There is evidence from ancient shorelines and coastal dunes that sea level fell more than 2 metres over the period from 4,000 to 2,500 years ago. Ancient harbour works around Naples and the Adriatic, dating from about 2,500 years ago and now submerged, suggests a mean sea level about 1 metre below the present. However, tectonic movement around the Mediterranean Sea may confound this conclusion. There is more confidence that world sea level rose during Roman times reaching a level, about 400 AD, which is comparable to or slightly higher than the present.

The archaeological and cultural evidence pointing towards a highly variable climate through the Holocene up to Roman times is compelling. In areas of human settlement from which data are available there does appear to be a degree of synchronicity between events in different regions. Glacial advances and abnormally cold and wet weather over Europe were often contemporary with drier conditions across the subtropics. The rise and fall of sea level by about 2 metres suggests significant water becoming locked into the polar ice sheets and mountain glaciers. Very clearly, many of the relatively abrupt climate changes had devastating impact on local populations who had neither the resources nor technology to quickly adapt to the changed climate.

Explanations for Climate Variability

A belief that carbon dioxide and its radiative properties may be important in regulating climate and its variability is not new. In 1859, the Irish-born John Tyndall began studying the radiative properties of various gases[31]. Tyndall constructed the first ratio spectrometer to measure the absorptive power of gases, including water vapour and carbon dioxide, and identified significant differences in the abilities of various gases to absorb and transmit radiant energy. He showed that molecules of water vapour, carbon dioxide and ozone are the best absorbers of heat radiation, and that even in small quantities these gases absorb much more strongly than the atmosphere itself. Tyndall concluded that, among the constituents of the atmosphere, water vapour is the strongest absorber of radiant heat and is therefore the most important gas controlling the surface temperature of the earth. Without water vapour, he concluded, the earth's surface would be "held fast in the iron grip of frost". He later speculated on how fluctuations in water vapour and carbon dioxide could be related to climate change.

The glacial epochs
The Swedish chemist, Svante Arrhenius (a winner of the Nobel Prize for his work on the rate of chemical reactions) attempted to quantify the impact of fluctuating concentrations of atmospheric carbon dioxide on the earth's climate. Palaeontologic evidence had by then accumulated suggesting that the earth suffered a series of severe glacial periods during the past few million years. In the report summarising his calculations and conclusions[32], Arrhenius noted that

---

31 NASA Earth Observatory Library, www.earthobservatory.nasa.gov
32 *On the influence of carbonic acid in the air upon the temperature of the ground.* Reprinted in English by the Royal Swedish Academy of Sciences in a commemorative volume, The legacy of Svante Arrhenius: understanding the greenhouse effect, 1998.

"temperature in the Arctic zones appears to have exceeded the present temperature by about 8 or 9 degrees" and "When the ice age had its greatest extent, the countries that now enjoy the highest civilisation were covered in ice. This was the case with Ireland, Britain (except a small part in the south), Holland, Denmark, Sweden and Norway, Russia (to Kiev, Orel, and Nijni Novgorad), Germany and Austria (to the Harz, Erz-Gebirge, Dresden, and Cracow). At the same time an ice cap from the Alps covered Switzerland, parts of France, Bavaria south of the Danube, the Tyrol, Styria, and other Austrian countries, and descended into the northern part of Italy. Simultaneously, too, North America was covered in ice on the west coast to the 47th parallel, on the east coast to the 40th, and in the central part to the 37th (confluence of the Mississippi and Ohio rivers). In the most different parts of the world, too we have found traces of a great ice age..."

Eccentricity
The earth follows an elliptical path around the sun and completes one orbit each year.
The eccentricity relates to the distances of the earth and sun from the centre of the ellipse
when the earth is at its farthest distance from the sun (aphelion). Eccentricity changes from
near circular to being highly elliptical with a period of nearly 100,000 years. Currently the orbit
is near circular.
The earth intercepts less solar radiation when it is at aphelion than when it is at perihelion. The
total amount of solar radiation intercepted over a year changes very little with changing
eccentricity.

Obliquity
The earth's obliquity is the angle between the axis of rotation (the north-south axis) and the
vertical to the plane of the earth's orbit around the sun.
The earth's obliquity varies between 22°2' and 24°20' (at present 23°27') with a period near
40,000 years.
The amount of solar radiation intercepted by the earth does not change with obliquity but at
maximum obliquity more solar radiation is
received over middle and higher latitudes
of the summer hemisphere and winter
radiation is reduced.

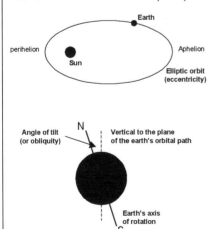

Precession
The earth's precession is the slow rotation
of the north-south axis around the vertical
to the plane of the earth's orbit around the
sun.
The precession period is about 20,000
years.
Currently, the earth is closest to the sun in
early January and the summer insolation
received over the southern hemisphere is
slightly more than that received over the
northern hemisphere.

Figure 6:   The earth's orbital characteristics

From his calculations, Arrhenius concluded that the temperature of the Arctic regions would rise by about 8–9°C if the concentration of carbon dioxide (carbonic acid) in the air increased 2.5–3 times above its then value. For the other extreme, in order to get the temperature necessary for an ice age between the 40th and 50th parallels (lowering the temperature 4–5°C) the concentration of carbon dioxide would have to sink to 0.62–0.55 of its then value. Arrhenius also concluded from his calculations that the whole earth should have undergone about the same variations in temperature so that "genial or glacial epochs must have occurred simultaneously on the whole earth".

Arrhenius addressed the question of whether such changes in atmospheric carbon dioxide were feasible. He suggested that volcanic 'exhalations' are the chief source of carbon dioxide for the atmosphere and that the amount of carbon dioxide in the air is insignificant when compared to the carbon fixed in the earth's coal and limestone formations. However, in the formation of coal and limestone, carbon must have had residence in the atmosphere as carbon dioxide. Arrhenius concluded that it is highly unlikely, therefore, that the rate of formation and uptake of atmospheric carbon dioxide should be in balance. As a consequence, variations in concentration necessary to explain the past variations of climate are feasible.

Alternative astronomical theories to explain climate variability and the formation of ice ages had also had their adherents in the later half of the 19th century. The German mathematician Johannes Kepler had discovered, early in the 17th century, that the orbits of the planets around the sun were elliptical. By the middle of the 19th century it was also known that the earth's axis of rotation, which is tilted at 23.5 degrees, precessed through 360 degrees over a period of about 22,000 years. Thus every 22,000 years the earth is closest to the sun at northern hemisphere midsummer and 11,000 years later it is closest at northern hemisphere midwinter. These variations change the relative amounts of summer solar radiation received by the two hemispheres. For example, currently the earth is closest to the sun on 3 January and furthest on 4 July and the southern hemisphere receives more summertime solar radiation at the top of the atmosphere each year than the northern hemisphere.

In 1842 the French mathematician Joseph Adhemar published 'Revolutions of the Sea', a theory based on the precession and elliptical orbit of earth. He proposed a 22,000-year recurrence of ice ages, with the northern hemisphere being out of phase with the southern hemisphere[33]. The German scientist Alexander von Humboldt pointed

---

33 See John Gribbin, Science: A History. The Penguin Press, 2002, for a discussion of the theory and why it was incorrect.

out a deficiency in the theory, in 1852. Von Humboldt noted that the idea of one hemisphere getting warmer while the other is getting colder is wrong because the wintertime deficiency of solar radiation in one hemisphere is balanced by summer surplus. Over the course of a year each hemisphere receives the same solar radiation at the top of the atmosphere.

The next serious astronomical theory was proposed in 1864[34] by James Croll, a Scot. Croll had access to very laborious calculations that detailed the changing ellipticity of the earth's orbit around the sun with time. According to Croll's model, based on the changes in ellipticity and precession, alternating ice ages would occur in each hemisphere and would be embedded in an ice epoch hundreds of thousands of years long. He estimated the most recent ice epoch to have commenced about 250,000 years ago and was followed by a warm interval that commenced about 80,000 years ago. In later publications Croll alluded to the importance of the changing tilt of the earth's axis.

Croll's astronomical theory lost support as geological evidence accumulated during the second half of the 19th century and pointed to the most recent ice age ending about 15,000 years ago, not 80,000 years ago. The conflicting geological evidence is cited by Arrhenius who noted, "the great advantage which Croll's hypothesis promised to geologists, viz. of giving them a natural chronology, predisposed them in favour of its acceptance. But this circumstance, which at first appeared advantageous, seems with the advance of investigation rather to militate against the theory, because it becomes more impossible to reconcile the chronology demanded by Croll's hypothesis with the facts of observation."

The astronomical basis for climate variability regained some support from the work of the Serbian mathematician Milutin Milankovitch[35]. In 1920, Milankovitch published a book giving a mathematical description of the climates of Earth, Venus and Mars, including astronomical calculations demonstrating that variations in the earth's ellipticity, precession and obliquity (or tilt) could alter the amount of heat arriving at different latitudes.

Milankovitch concluded that the variations in ellipticity, tilt and precession were sufficient to vary the amount of solar radiation received at high latitudes and cause ice ages. The pioneer climatologist Vladimir Koppen suggested that it was summer temperatures over the high latitudes that were important in the development of ice ages[36]. Koppen's reasoning was that it is always cold enough for snow to fall

34   John Gribbin, Ibid.
35   NASA Earth Observatory Library, Ibid.
36   John Gribbin, Ibid.

over the high latitudes in winter, so what matters is the temperature of summer and the degree to which accumulated snow stays unmelted. Milankovitch carried out further detailed calculations that provided estimations of temperature variations at latitudes 55, 60 and 65 degrees North and got what seemed to be a very good match between the astronomical rhythms and the then available geological evidence of past ice ages.

Milankovitch's theory on astronomical cycles seemed to be vindicated when polar ice cores were recovered, firstly from Greenland then from Antarctica, and provided better resolution and timing of past climate variations. The first drilling into polar ice sheets was accomplished in 1956 in northwest Greenland. However, it was not until 1966 that the first ice core to reach bedrock was recovered from Greenland. A core from the Greenland Summit, drilled as part of the European Greenland Ice Core Project, reached 3028 metres before striking bedrock. The changing climatic conditions are recorded in the annual layers of snow that accumulated over the polar regions and so a deep core is effectively a record of the climatic history. From the ice cores, and their changing chemical and physical composition, it was possible to reconstruct temperature variations back over the past 450,000 years. The periods of climatic variation that were identified corresponded to the primary astronomical cycles of precession (20,000 years), obliquity (40,000 years) and eccentricity (100,000 years).

Despite the very good statistical correlation between seasonal variation of solar irradiance and the reconstructed climate record there are still unanswered questions as to how these relatively small changes are amplified into major climatic variations. In particular, the major ice ages have a lifecycle of about 100,000 years and this corresponds to the eccentricity period. The ice ages are at their maximum when the earth's orbit is highly elliptical and the total daily solar radiation received at the top of the atmosphere is 20 to 30 percent greater at perihelion (position in the orbit closest point to the sun) than at aphelion (furthest point). However, the total annual solar radiation received at the top of the atmosphere varies little with eccentricity of the orbit despite the significant seasonal variation with changing ellipticity. Currently, the earth's orbit is almost circular and the seasonal variation of daily solar radiation at the top of the atmosphere is about 5 percent.

The large ice sheets of the last glacial epoch began melting some 15,000 years ago. The global temperature rose to reach a maximum about 8,500 years ago although the temperature rise has not been regular.

The earth's emergence from the last ice age is consistent with the Milankovitch cycles theory and the current reduced eccentricity of the earth's orbit. However, the underlying thermodynamic processes of the climate system associated with the warming are poorly understood. Questions arise about what was the trigger for the onset of a global warming trend and why are the warm interglacial epochs relatively short when compared to the glacial epochs.

Although Milankovitch's astronomical cycles are suggested as the basis for major climate variation, the internal processes of the climate system that respond and amplify the climate response are not resolved. There is even the suggestion that only a small amount of the climate variance (less than 20 percent) can be attributed to solar forcing related to the Milankovitch cycles[37]. Analysis of palaeoclimate records indicates that there is only small energy excess in the obliquity (40,000 yr) and precessional (22,000 yr) bands and energy in the eccentricity (100,000 yr) band is indistinguishable from a broadband stochastic process. Thus, support for the view that Milankovitch forcing somehow controls the records is weak because processes indistinguishable from stochastic dominate the records.

Recent global warming

A critical issue in the climate change debate is the extent to which near-surface air temperatures have varied over the past millennium, and whether warming over the 20th century should be considered as unusual. The apparent variability of climate on geological timescales, including long periods of glaciation and intervening warm epochs, would suggest that recent warming is probably not unusual. Indeed, estimates based on data from the deep ice cores of Antarctica suggest that temperatures were warmer in the early part of the Holocene 4–8,000 years ago. Also, the global temperature of the Holocene is not unusual when compared to previous interglacial periods.

New reconstructions of climate patterns have led the IPCC to conclude that, at least for the northern hemisphere, the temperature in the 20th century is likely to have been the warmest of any century during the last 1,000 years and the 1990s were the warmest decade. The fact that there is so much controversy over the issue only highlights its importance and the paucity of compelling data one way or the other. The lack of widespread instrumental data prior to the mid-19th century has allowed the debate over the significance of recent global warming to flourish.

The relatively short length of instrumental records precludes quantitative statements with any certitude about the variability of regional or global climate. Even assessments for recent times, using modern instruments, suffer because large parts of the globe continue

---

37 Wunsch, C, 2003: Quantitative estimate of the Milankovitch-forced contribution to observed climate change. (Unpublished manuscript)

to be poorly sampled. Over the oceans sparse ship observations are supplemented by estimates of sea surface temperature based on satellite observations but these latter have only been reliably available for the past two decades. Many sparsely populated land regions continue to have limited numbers of observing sites, and civil unrest and warfare have interrupted the record in many areas. The very fact that most meteorological observing sites are located in or close to urban areas opens up the potential that the expanding areas of paving and buildings are contributing significantly to the warming trend that is being observed (see Figure 3).

The absence of widespread instrumental records before the middle of the 19th century has necessitated the use of proxy data to reconstruct past climates. The use of proxy indicators as the basis for climatic reconstructions is linked to the classification of regional climate according to vegetation type. The pioneer in this work was the Russian born climatologist Vladimir Koppen who, during the late 19th century at the Hamburg Naval Observatory, assembled climate data from around the world. By the early 20th century he had analysed the data and produced maps of climate classification related to five principal vegetation types. The assumption is that each vegetation type is adapted to its own climate regime. Therefore, if there is evidence that the vegetation type has changed then it is logical that the climate has also been different in the past.

When it comes to interpreting proxy data for quantitative trends we have a much more complex issue. Invariably a proxy responds to more than one environmental factor and its record does not encompass the full variability of the local climate indicator to which it is being linked. For example, the width and density of tree rings represent annual growth patterns that respond to a combination of temperature, soil moisture, soil nutrients, sunlight and carbon dioxide. A deviation of the seasonal cycle from average (whether warmer or colder, wetter or drier) will cause a change in the annual growth ring. Warmer and wetter conditions are generally more favourable to growth than colder and drier conditions and tropical forests have more luxuriant growth than the forests of high mountain or polar regions. By carefully selecting the location and species it is possible to identify a statistical relationship between the annual growth pattern (whether tree ring width or density) and temperature.

It must be recognised, however, that the annual growth rings cannot reproduce the full fidelity of seasonal temperature variations. It is assumed that the rate of tree growth varies with average annual temperatures, or even seasonal temperatures where there is a dominant seasonal growth pattern. However, the average temperature over a period reflects the changing weather patterns and the fluctuations between colder and warmer episodes. Trees will respond to the fluctuations and an assumption of a linear relationship has only limited validity. In particular, tree growth for individual species is well adapted within a temperature range but will become seriously curtailed as temperatures approach the limits of adaptation. The stability of the relationship between the proxy and the regional climatic indicator (whether temperature or precipitation) that it is meant to represent is important for a range of biologically based proxies.

Other proxies identify when the climate was on either side of a threshold. For example, the existence of dust accumulation in a sediment core is a likely indicator of aridity but we know neither the extent of aridity under these conditions nor the magnitude of precipitation when vegetation binds the soil and dust accumulation is absent. Similarly in ocean cores, the changing ratio of skeletal remains from different species will point to a change of surface water temperatures, but not the magnitude either way. At times there will be a preponderance of marine life forms that occupy warmer surface waters but at other times the preponderance is of those that occupy colder surface waters. From the record of changing ratios it can be deduced that the temperature of the ocean surface water has changed, when it changed and, perhaps, the threshold temperature of change. However, as with tree rings, the full range of temperature change is rarely captured. These types of data identify the timing of past climatic variations and yield important qualitative information about the magnitudes of the variations.

We have earlier referred to reconstructions of climate from earlier times derived from historical accounts and archaeological information. These reconstructions, mainly for Europe and the Middle East but also including China and the Americas, identified major changes from present climatic conditions. Hubert Lamb's account of the centuries leading up to and covering the last millennium continues in the vein of fluctuating climate and human impacts. Historical and cultural information for the recent period is more pervasive and includes direct accounts of devastating cold, paintings depicting rivers and lakes that were then frozen but are now ice-free, and paintings of mountain glaciers identifying whether there had been an advance or retreat from the present. Lamb has concluded that Europe had been relatively warm during the medieval period (approximately 800 to 1300) but then plunged into a period of cold that lasted until the mid 1800s.

The Medieval Warm Period and Little Ice Age, as these epochs have been dubbed, became widely accepted as real climate shifts affecting at least the Europe and the North Atlantic Ocean region. Evidence of contemporaneous advancing mountain glaciers in various parts of the world has been taken to suggest that the period of cooling, and by inference the earlier warm epoch, were of global dimensions. During the warmer epoch, Iceland and Greenland were colonised and the cultivation of a range of domestic plants was extended far poleward from the present limits. However there is an absence of cultural and historical data to corroborate that the warm epoch was global. Indeed, there is some evidence that China and Japan were abnormally cold during the European medieval warm period.

Statistical methods for climate reconstruction

Over the past three decades the amount and variety of proxy data has increased rapidly from widely dispersed locations around the globe. The data have come from tree rings, coral cores, ocean and lake sediments and ice cores from polar and mountain glaciers. Each proxy gives a record of a unique aspect of climate for its source region but no proxy type is able to provide global coverage. Corals are restricted to tropical waters, ice cores are limited to polar and mountainous regions and trees respond differently to climate variability depending on the species, its location and its prevailing climatic regime. The real

Reconstruction of annual northern hemisphere temperature trend from proxy data (tree rings, coral cores and historical records). The heavy dark line is the 50-year average. The grey region is the 'confidence' in the reconstruction, which degenerates farther back in time because of the lesser number of proxies contributing to the reconstruction. The reconstruction for the 20th century is a correlation between the full set of proxies and the global temperature field (the training period).

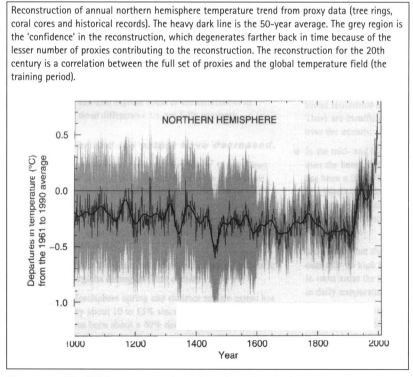

Figure 7:   The IPCC reconstruction of northern hemisphere surface temperature over the past millennium

challenge has been to integrate these data to give a coherent global picture of how climate has changed in the past. Interpreting these proxy climate data has produced one of the major controversies of the global warming issue.

An innovative technique for statistical analysis combines a range of proxy indicators to reconstruct regional and hemispheric annual temperatures from the past. The multi-proxy temperature reconstructions rely on annually resolved proxies that reflect temperature variability and are calibrated with the instrumental record of the 20th century. The calibration period provides a basis for extending the temperature record back in time over the much longer period for which the proxy records are available. One such reconstruction by Michael Mann, then of the University of Massachusetts, USA, and colleagues has been used by the IPCC as the basis for interpreting the global surface temperature record of the past millennium[38,39]. This reconstruction, now widely referred to as the 'hockey stick' because of the steadily declining temperature trend for the first 900 years followed by a steep rise in the 20th century, features large in the IPCC conclusions on anthropogenic global warming.

The hockey stick is controversial because it is counter to previous beliefs about the climate record derived largely from cultural and historical accounts and their interpretations. The temperature reconstruction does not identify a Medieval Warm Period or a Little Ice Age pattern. The IPCC claim, based on the temperature reconstructions from proxy records, is that the Medieval Warm Period and the subsequent Little Ice Age are regional anomalies and not identifiable global phenomena. Moreover, the IPCC makes the claim that the reconstructed temperature record shows that the 20th century was the warmest of the millennium. A close look at the methodology used to produce the hockey stick profile suggests that the IPCC has drawn conclusions that are not justified from the data and methodology.

The rudiments of the reconstruction methodology are the following. Firstly, the set of global fields of annual mean near-surface air temperature from the period of the instrumental record can be represented by a set of Principal Components[40]. Secondly, because the

38 Mann, M.E., R.E. Bradley and M.K. Hughes, 1998. Global-scale temperature patterns and climate forcing over the past six centuries. Nature vol 392, pp779–787.

39 Mann, M.E., R.E. Bradley and M.K. Hughes, 1999. Northern hemisphere temperatures during the past millennium: inferences, uncertainties and limitations. Geophys. Res. Lett., vol 26 No 6, pp759–762.

40 A Principal Component is a set of numbers over the period that represents one recurring aspect of the pattern variability and contributes to the overall pattern variance. If the set of Principal Components is representative of recurring component patterns the set can be used to reconstruct the changing pattern over time.

signal of each proxy is assumed to vary systematically with the fluctuations of local climate, there is a unique relationship between the Principal Components and the chosen set of proxy records. Through statistical analysis over a 'training' period (part of the instrumental record when global temperatures are relatively well known) this relationship can be identified. Thirdly, the unique relationship and the full set of proxy records are used to reconstruct a time history of each of the Principal Components of near-surface air temperature. The set of derived Principal Components extending back in time over the period when proxies are available provide the basis for reconstructing near-surface temperature fields over the same extended period.

Principal Components of climatological fields have been extensively used to deconstruct the variability of the fields into their dominant patterns. These dominant patterns, or principal components, give insights into the underlying causes of variability. The first Principal Component of global near-surface air temperature is related to the variation of the global mean and explains the largest proportion of the variance. The second Principal Component explains the next largest proportion of the variance and has a characteristic El Niño pattern; it represents interannual spatial variability of temperature associated with El Niño-Southern Oscillation forcing. Subsequent Principal Components explain decreasing proportions of the interannual and spatial variability.

The temperature fields used for the 'training period' of the reconstruction are widely recognised analyses of near-surface air temperature and sea surface temperature fields[41] that cover the period 1902–1995 AD. These fields are the basis for identifying a set of Principal Components. Over the 'training period' 1902–1980, the first Principal Component explains 73 percent of the variance associated with the northern hemisphere annual mean temperature. The combined first five principal components explain 85 percent of the variance. Subsequent addition of more principal components increases the percentage of variance that is explained.

For the surface temperature reconstruction, a total of 112 proxies were selected to be the predictors of annual 'surface temperature'[42]

41   Jones, P.D. and K.R. Briffa, 1992. Global surface air temperature variations during the 20th century: Part I – spatial, temporal and seasonal details. Holocene 1, pp165–179.
42   The global surface temperature data set used for calibration was composed of near-surface air temperature (ie, from standard meteorological screens) over land and sea surface temperature over the oceans but such distinctions are lost in the analysis. However the distinction is not trivial because there is often a significant difference between the surface temperature (whether over land or sea) and near-surface air temperature. Radiation budget calculations derive the surface temperature, not the near-surface air temperature.

over the last millennium. The set of proxies was made up of dendroclimatic (both tree ring width and density), ice core, ice melt, coral and long instrumental records that provided annual resolution. It is claimed that the mutual information contained in a diverse and widely distributed set of independent climatic indicators (the multi-proxy data set) can more faithfully capture the consistent climate signal that is present, thus reducing the compromising effects of biases and weaknesses inherent with individual indicators.

The relationship between the set of Principal Components and the set of proxies was identified over the training period, 1902–1980 AD. The solution is a matrix of coefficients relating the different proxies to their closest linear combination of the set of Principal Components. The dimensions of the matrix are the number of principal components and the number of proxies. The principal components to be used were the group that maximised the explained variance over the calibration interval and this was 11 for the 112 proxies available over the training period. The solution matrix is then used to reconstruct the Principal Components of the global surface temperature field over the training period, or any other period for which the proxy data are available.

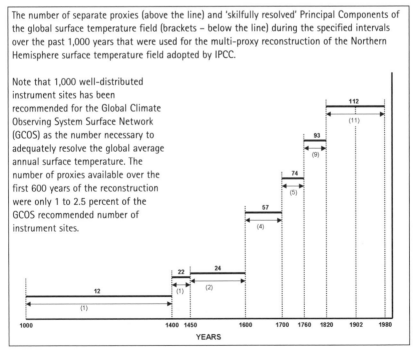

Figure 8:   The number of proxies contributing to the reconstruction of the northern hemisphere temperature trend adopted by IPCC

A critical deficiency in the multi-proxy methodology is that not all proxy data are available across the full period of the last millennium. Also, as the number of proxies decreases so too the number of Principal Components that can be resolved also decreases. Thus the ability of the solution matrix to recreate the variance during the training period degrades as proxies are omitted from the predictor set. For example, the 112 proxies and 11 retained Principal Components used after 1820 AD are claimed to recreate 76 percent of the northern hemisphere surface temperature variance over the training period (1902–1980). For the period after 1600 AD there are still 57 proxies available and 4 retained Principal Components that can potentially recreate 67 percent of the training period variance. Prior to 1600 AD, however, less than 30 proxies are available and these, with reduced retained Principal Components, can potentially only reconstruct about 40 percent of the variance of northern hemisphere surface temperature during the training period. The number of proxies that were available and the number of Principal Components used for each interval of the reconstruction are shown in Figure 8.

The reconstruction of northern hemisphere surface temperature prior to 1400 AD is compromised by the nature of the 12 proxies that are used. Three of the proxies are the first, second and third Principal Components of a data set of North American tree ring widths. One is a data set of tree ring widths from Tasmania (lat. 43°S). One is a data set of tree ring widths from northern Patagonia (lat. 38°S). Two are from the same Quelccaya (lat. 14°S) ice core – ice accumulation and oxygen isotope respectively. Of the nominal 12 independent proxies four are from the southern hemisphere and three others represent the same data set. Only 6 separate data sets represent the northern hemisphere.

In the period prior to 1450 (nearly half of the length of the surface temperature reconstruction) only the first Principal Component of the surface temperature field can be skilfully resolved. As noted earlier, the first Principal Component contributes to the spatial mean but is not the sole contributor. There is no expectation that the reconstruction for this early period will explain more than 40 percent of the variance of the actual northern hemisphere mean surface temperature because this is the variance that the first Principal Component and all the relevant proxies explain during the calibration period. Overall, as more proxies and Principal Components become available during the millennium, the reconstructed values should better represent the actual northern hemisphere surface temperature. This limitation of the methodology means that any significant hemispheric mean surface temperature variation that actually occurred in the earlier part of the millennium, whether it is on interannual or longer timescales, would be significantly dampened in the reconstructed record.

It is surprising, therefore, given the limitations in the methodology, that the IPCC chose to use the analysis in support of its claim that the increase in temperature in the 20th century is likely to have been the largest of any century during the past 1,000 years. It is even more surprising when, in presenting their findings, the authors of the analysis had concluded, "Our reconstruction thus supports the notion of relatively warm [northern] hemispheric conditions earlier in the millennium, while cooling following the 14th century could be viewed as the initial onset of the Little Ice Age..."[43]. The author's conclusion is consistent with an earlier analysis of multi-proxy data and a finding of natural oscillations, including "...a global-scale transition from warmer conditions before 1500 to the colder conditions of the sixteenth to eighteenth centuries"[44].

The IPCC conclusions, based on the multi-proxy temperature reconstructions, must be considered as unreliable because the limited proxy data available from early in the millennium do not reconstruct annual mean surface temperatures of the northern hemisphere with the same fidelity as instrumental observations available from the mid-19th century.

The IPCC denial of the Medieval Warm Period as either an hemispheric or global event has been received with scepticism by many scientists familiar with proxy climate data. In response to the multi-proxy reconstruction and its adoption by the IPCC, Wallace Broecker of Columbia University, USA posed the question, "Was the Medieval Warm Period global?"[45]. Broecker broadly reviewed the available proxy data and its value in reconstructing past climates. His judgement was that only two proxy types, the elevation of mountain snowlines and borehole thermometry, can yield temperature reconstructions that are accurate to 0.5°C, and thus suitable for resolving the fluctuations of the Holocene period.

Around the world, except for Antarctica, advancing mountain glaciers reached their maximum extent at about 1860 AD. The well-documented evidence of glacier extent that is available strongly suggests a globally synchronous cooling during the Little Ice Age period and a globally synchronous warming since about 1860 AD. Historical evidence[46] from the Swiss Alps corroborates what has already been described for Europe and the North Atlantic Ocean and

43 Mann, R.E., R.E. Bradley and M.K. Hughes, 1999. Ibid, p762.
44 Mann, M.E., J. Park and R.S. Bradley, 1995. Global interdecadal and Century-scale climate oscillations during the past five centuries. Nature, vol 378, pp266–370 (p 269)
45 Broecker, W.S., 2001. Was the Medieval Warm Period warm? Science, vol 291 pp1497–1499.
46 H. Holzhauser, cited in Broeker (2001), Ibid.

points to a significantly warmer period preceding the Little Ice Age. The warmer conditions over the Swiss Alps can be inferred from the construction of a wooden aqueduct across the valley below the Grosser Aletsch Glacier that has been dated at about 1200 AD. The aqueduct was partially destroyed in 1240 AD during an advance of the glacier and had to be totally re-routed in 1370 AD after a further advance. Radio-carbon dating of wood and peat fragments being disgorged from beneath a number of Swiss glaciers suggests earlier warm periods about 1,500, 2,400, 4,300, 6,600 and 8,700 years ago. The near global synchronicity of glacial advances and retreats over the past millennium points to a global forcing mechanism for climate warming and cooling.

The thermal profiles in boreholes can be mathematically analysed to reconstruct a smoothed temperature record at the surface. However, only the broad features of the time history are captured. Boreholes from the polar ice caps of Greenland and Antarctica clearly reflect the colder temperatures during the last glacial period. A fluctuation of the Greenland profile in the late Holocene matches the timing of the swing from the Medieval Warm Period to the Little Ice Age. Analysis of 6,000 continental borehole records[47] from around the world suggests that 500 to 1,000 years ago temperatures were warmer than today and about 200 years ago they cooled to a minimum some 0.2°C to 0.7°C below present. Again, the highly smoothed reconstructions are indicative but do not prove the existence of a globally coherent shift from a Medieval Warm Period to a Little Ice Age.

Willie Soon and Sallie Baliunas of the Harvard-Smithsonian Center for Astrophysics, USA have considered an array of proxy data records to represent an ensemble of expert opinions to test whether the Medieval Warm Period and Little Ice Age were widespread climate anomalies[48]. They addressed individual proxy records with three seemingly simple questions:

(i)   Is there an objectively discernible climatic anomaly during the Little Ice Age interval, defined as 1300–1900 AD?

(ii) Is there an objectively discernible climatic anomaly during the Medieval Warm Period, defined as 800–1300 AD?

(iii) Is there an objectively discernible climatic anomaly during the 20th century that is the most extreme (the warmest if such information is retrievable) period in the record?

---

47   S. Huang et al, cited in Broeker (2001), Ibid.

48   Soon, W. and S. Balliunas, 2003. Proxy climate and environmental changes of the past 1000 years. Climate Research, vol 23, pp89–110.

The results of the study have been subjected to criticism, not only because of the nature of the findings but also because of the definition of what constitutes an anomaly. According to Soon and Baliunas, "Anomaly is simply defined as a period of more than 50 years of sustained warmth, wetness or dryness, within the stipulated interval of the Medieval Warm Period, or a 50 year or longer period of cold, dryness or wetness within the stipulated Little Ice Age". The definition is further refined by the proviso that, "the terms Medieval Warm Period and Little Ice Age should indicate persistent but not necessarily constant warming or cooling, respectively, over broad areas".

For many critics, the nature of the definition means that the interpretation of a complex record is open to ambiguity. Soon and Baliunas were certainly aware of the complexity of the climate record. They noted that the notion of a Medieval Warm Period or Little Ice Age with sharply defined transitions may be convenient but it is not realistic because of large regional differences in the timing of both phenomena. To this we should add the confounding influence of multi-decadal to centennial scale variability that appears to be present in many proxy records.

The Soon and Baliunas review is an assessment of 141 published studies of local regional and global representations of climate indicators that extend over the past 1,000 years. In some cases the individual researchers had themselves identified a Medieval Warm Period or Little Ice Age in their data; in others the proxy record had to be addressed independently in terms of the 3 questions. From their ensemble of answers to the three questions, Soon and Baliunas concluded that the Little Ice Age and Medieval Warm Period were widespread and nearly synchronous phenomena. However, in most of the proxy records the 20th century does not contain the warmest anomaly of the past millennium. Moreover, the proxies show that the 20th century is not unusually warm or extreme. Here we should add a note of clarification: Soon and Baliunas were comparing the proxy records of the 20th century, not the instrumental temperature record.

Stephen McIntyre and Ross McKitrick of Canada have challenged the validity of the IPCC 'hockey stick' surface temperature reconstruction[49]. In attempting to replicate the computation they identified what is claimed as "collation errors, unjustifiable truncation or extrapolation of source data, obsolete data, geographical location errors, incorrect calculation of Principal Components and other quality control defects". Using the allegedly erroneous multi-proxy database McIntyre and McKitrick were able to essentially reproduce the 'hockey

49 McIntyre, S., and R. McKitrick, 2003. Corrections to the Mann et al (1998) proxy database and northern hemispheric average temperature series. Energy and Environment, vol 14, No 6, pp 751–771.

stick' profile. Using corrected and updated source data McIntyre and McKitrick found that surface temperatures in the early 15th century exceeded any values in the 20th century. They concluded that the 'hockey stick' temperature profile is primarily an artefact of poor data handling, obsolete data and incorrect calculations of Principal Components. To this should be added the probable sensitivity of the methodology that is a consequence of so few proxies with annual resolution that are available prior to the 16th century.

In summary, the different tools available for estimating temperature fluctuations during the past millennium do not permit a quantitative assessment of how local, regional, hemispheric or global temperatures have varied. The regular daily meteorological observations from around the globe have provided seasonal and annual estimates with a high degree of fidelity for approximately 140 years. Unfortunately, changing instrumentation and environmental factors, especially expanding urbanisation, contaminate these data. Nevertheless, there is a clear signal of warming during the 20th century and the pattern is relatively consistent over the tropics and both polar regions.

A number of physical, chemical and biological indicators, with millennium length records, have been used as proxies for climate. Rarely do these provide a one-to-one correspondence with a single climate indicator although many have been used as proxies for temperature. The individual proxy records display significant variability over the last millennium and suggest that the earlier centuries were warmer while the climate was generally cooler between the 16th and 18th centuries. The proxy records cannot be directly compared with the later instrumental records because the characteristic response of proxies to temperature change is generally not linear. In particular, biologically based proxy indicators are generally only responsive within the temperature range to which they are adapted. The cumulative evidence is that the proxy data are generally consistent with the glacial advance as the Medieval Warm period gave way to Little Ice Age conditions. These phenomena are well documented for the European-North Atlantic region, especially in sub-polar latitudes, but strong evidence for a worldwide climatic response is more lacking than contradictory.

The global surface temperature reconstruction for the past millennium adopted by the IPCC must be treated with caution, if not scepticism. Firstly, an aggregation of proxies that respond differently to climate is expected to reduce the sensitivity of the methodology to real climate variations. Secondly, an expectation that the use of less than 25 proxies early in the millennium can truly reflect climate variability is not realistic (modern thinking encapsulated in the Global

Climate Observing System (GCOS) Surface Network is that at least 1,000 instrumented and well spaced observing sites is a minimum network to adequately monitor global average surface temperature variations[50]). Thirdly, there is need for an independent audit of the temperature reconstruction in light of the allegation of errors in the proxy database construction. The claim made by the IPCC that the 20th century was the warmest of the millennium is not substantiated. In addition, the IPCC claim is clearly at odds with cultural experience and raw proxy records from many regions of the globe.

---

50   Peterson, T., H. Daan and P. Jones, 1997. Initial selection of a GCOS surface network. GCOS-34.

Processes of the Climate System

The simple one-dimensional characterisation of the climate system portrayed by the IPCC (see Figure 1) is grossly inadequate. It completely ignores fundamental processes that regulate energy flow through the climate system and has led to erroneous conclusions about the importance of radiative forcing of the climate system. There are many important processes in addition to radiation that also need to be considered because of their roles in regulating energy flow through the climate system and their contributions to climate variability. What we need to recognise is that the climate system is multi-dimensional beyond solar and terrestrial radiation processes.

The earth's climate system is exceedingly complex. In its simplest it consists of the land surfaces, the oceans, the polar and high mountain ice sheets, and the enveloping atmosphere. For practical purposes we treat the land geography as fixed (ie, neglecting continental drift). However, the oceans and atmosphere are in constant motion and the ice sheets expand and contract, both seasonally and on longer timescales. Clouds are constantly forming and dispersing and are crucial for the climate system as they release latent energy of water vapour to the atmosphere. Clouds also locally affect flows of radiant energy. The biosphere of the land surface is in constant change, especially as vegetation responds to the seasonal cycles and longer period variations of climate. The biosphere is particularly important in considerations of surface evapotranspiration within the hydrological cycle, and for plant photosynthesis and oxidation as part of the carbon cycle. Local salinity, and hence density, of the ocean surface layer is constantly being modified by imbalances between evaporation and precipitation over the open oceans. Surface runoff is also an important regulator of the surface water salinity of semi-enclosed seas.

The energy cycle, the hydrological cycle and the carbon cycle are crucially important components of the climate system. The energy cycle, in its simplest depiction, covers the absorption of solar radiation, the exchange of energy between the components of the climate system, and the emission of heat to space in the form of longwave radiation. The basic hydrological cycle involves evapotranspiration from plants and evaporation of moisture from ocean and lake surfaces, the condensation of water vapour in clouds, and the precipitation of rain and snow at the surface. An important component of the carbon cycle is uptake of carbon dioxide by plants during photosynthesis and later oxidation of decaying plant material that returns carbon dioxide to the atmosphere. The absorption of carbon dioxide by cold surface water over high latitudes and its release to the atmosphere from upwelling water in the tropics is also an important component of the carbon cycle.

The energy cycle is closely tied to variability of the climate system. Almost nowhere is there energy balance at the top of the atmosphere. When averaged over a year there is net radiation excess over the tropics and radiation deficit over the polar regions. Therefore, there is a need for the circulations of the atmosphere and oceans to transport energy from the tropics to the poles. It is the local and regional imbalances of solar and terrestrial radiation and the resulting circulations of the atmosphere and oceans transporting energy around the globe that determine the climate and its variability. It is grossly misleading to neglect the spatial radiation imbalances and the ocean and atmospheric circulations, as the IPCC has done, in any theoretical consideration of climate change.

Within the climate system, about 20 percent of incoming solar radiation at the top of the atmosphere is absorbed within the atmosphere layer by clouds, greenhouse gases and aerosols. Over the oceans, about 90 percent of solar radiation that arrives at the surface penetrates and is absorbed in the surface layer and the remainder is reflected. As a consequence, the surface layers of the tropical oceans are acting as an energy reservoir and constantly accumulating energy. However the tropical oceans are stratified below the surface mixing layer and there is only limited downward mixing of energy into the deep cold ocean depths. The surface layers of the tropical oceans can be likened to an energy reservoir, being constantly replenished from the sun. The mixed surface layer of the tropical oceans is a continuing source of energy to the overlying atmosphere as heat and moisture are exchanged across the sea-air interface.

We might expect a simple overturning circulation to prevail in the atmosphere with warm air rising over the tropical oceans and sinking again over the polar regions. This does not happen because of the earth's rotation and conservation of absolute angular momentum by the atmosphere. Atmospheric angular momentum is related to the rate of spin about the earth's North-South axis of rotation. As air moves to higher latitudes the distance to the axis of rotation decreases and the rate of spin, and the westerly wind speed, increases (see Figure 19). Westerly wind jetstreams are a persisting feature over the subtropics and they form because of the increasing westerly wind speed component of air that spreads poleward from the deep tropical convection. The strong zonal speed of the westerly jetstreams inhibits the poleward movement of air in the upper troposphere.

Large scale overturning of the atmosphere is confined to the tropics and subtropics where the Hadley Cells are characteristic of each hemisphere. Ascent is concentrated in the deep convective clouds of the equatorial and monsoon regions. Within the deep tropical convection, latent energy and heat are converted to potential energy as updraught air ascends buoyantly to the high troposphere and is cooled. The high-level outflow from the convective clouds spreads to the subtropics and gently subsides in clear air. Potential energy is converted to heat in the subsiding air of the Hadley Cell circulation. The deep tropical convective clouds and the overturning Hadley Cells are essential to convert heat and latent energy of the boundary layer to heat within the troposphere. Tropical precipitation is a biproduct of the energy conversion process.

Over middle and higher latitudes the atmosphere achieves poleward transport of energy and other properties by way of standing waves of the westerly wind belt (also known as Rossby Waves) and transient eddies (or weather systems) that form at preferred locations within the standing waves. The Rossby Waves are able to transport heat, moisture and momentum because their shape is not regular. The poleward directed flow is warmer, moister and faster than the equatorward directed flow.

The circulation of air over polar regions is as a vortex with many characteristics of the solid rotation of the underlying earth. Transport of heat, moisture and momentum by the Rossby Waves and weather systems is reduced in the vicinity of the Poles. During winter, net radiation loss from the troposphere to space is compensated by large-scale subsidence of the polar air mass that converts potential energy to heat. The relatively high atmospheric pressure at the surface is evidence of the very cold air mass over the polar regions.

It is no coincidence that the Rossby Waves and weather systems of the middle and higher latitudes and the polar high pressure systems are most intense in the winter hemisphere. It is during winter when the equator to pole energy imbalance at the top of the atmosphere is greatest. As a consequence, the pole to equator temperature gradient is also a maximum and the atmospheric circulation responds to enhance the rate of energy transport. Over the summer hemisphere, although the net energy loss at the top of the atmosphere is reversed, the sub-freezing temperatures of the polar ice sheets maintain the pole to equator temperature gradient and continue to stimulate poleward energy transport.

Water, when compared to the atmosphere, has a relatively high density and inertia causes ocean currents to have strong persistence. The ocean currents can be likened to the flywheels of the climate system. When there is change to the tropical sea surface temperature patterns, such as during El Niño events of the Pacific Ocean, there is global impact on the atmospheric circulation. Many regional climate extremes that persist for up to a year and have significant social and economic impacts are linked to El Niño events. The full impacts of slow changes within the ocean circulation are yet to be identified, mainly because data are lacking. However there are clear signals that herald the importance of the oceans to climate on regional and global scales.

The relatively warm Gulf Stream and the North Pacific Current transport energy eastward across the North Atlantic and North Pacific Oceans respectively and transfer heat and moisture to the overlying air. The ocean heat source and the prevailing westerly wind-flow combine to ensure that the winter climates of Western Europe and the Pacific coast of North America are milder than corresponding western ocean margins at similar latitudes that come under the influence of winds with continental origins.

The Southern Ocean forms a continuous latitudinal band of ocean, only constrained by the Drake Strait between Antarctica and South America. The Antarctic Circumpolar Current gains its momentum from the overlying surface westerly winds. Not only do the surface westerly winds impart momentum to the Antarctic Circumpolar Current but they also cause an equatorward drift in the surface flow[51]. As a consequence of the Ekman turning there is constant upwelling of water at near freezing temperatures on the southern margin of the Antarctic Circumpolar Current. The rate of upwelling is related to the strength of the westerly wind and has important consequences, at least for Antarctic climate.

The polar ice sheets play an important role as an energy buffer within the climate system. Although cold compared to the tropics and subtropics, the high latitudes and polar regions also emit longwave radiation to space throughout the year. The radiation deficit of winter

---

51 As wind blows over the ocean surface it sets up a drift current. However the drift current flows across the direction of the wind, to the left in the southern hemisphere and to the right in the northern hemisphere. The turning of the drift current is related to the rotation of the earth and is known as Ekman turning.

months is largely offset by poleward transport of energy from the tropics. Part of the energy transport is as latent energy of water vapour. When water vapour sublimates to snow crystals latent heat is released to the atmosphere. Accumulation of snow at the ground over winter represents latent heat of sublimation that is available to the climate system to compensate for the net radiation loss to space. During summer, excess solar radiation contributes to melting ice and snow that accumulated over winter.

The seasonal expansion and contraction of the polar ice mass is not a simple reflection of an annual energy balance at the top of the atmosphere. More than eight times as much latent energy is released when water vapour sublimates during snow formation than is absorbed when a similar mass of snow melts to runoff. An additional complexity is that between 70 and 80 percent of solar radiation reaching a snow surface is reflected, depending on whether the snow is old or fresh. Therefore, during summer only a small fraction of the solar radiation at the top of the atmosphere is available to melt accumulated snow. This is not the case over forests and exposed land surfaces that have low reflectivity and where the surfaces strongly absorb available solar radiation. Notwithstanding the details of the complex thermodynamics, the polar ice sheets are an energy buffer within the climate system. On all timescales they have the ability to expand if there is a net energy deficit and contract if there is an energy surplus.

This brief overview identifies some of the complexities of the ocean and atmospheric circulations and their interactions as they combine to transport energy around the globe. In particular we note:

- The seasonal and zonal radiation imbalances at the top of the atmosphere that require energy transport within the climate system;

- The surface layer of the tropical oceans is an energy reservoir that absorbs and accumulates solar radiation;

- The ability of the polar ice sheets to buffer the climate system against variability in poleward energy transport;

- The complex response of the atmospheric circulation (tropical Hadley Cells, middle latitude Rossby Waves and weather systems, high latitude subsidence) to the pole-equator temperature gradient; and

- The surface winds of the atmospheric circulation that impart momentum to the ocean surface currents and influence upwelling and overturning.

These are key processes when considering climate variability on all timescales. It is, therefore, unrealistic to treat the climate system as a simple radiation balance model as has been portrayed by the IPCC. There is no justification for the concept of radiation forcing as the primary driver of climate change.

We will elaborate in more detail on radiation and other processes of the atmosphere and the oceans that are important in the seasonal transport of energy from the tropics to high latitudes. In addition, some of the main processes by which the oceans and atmosphere interact through exchange of heat, moisture and momentum will be described. In combination with the energy reservoir of the tropical oceans and the energy buffer of the polar ice sheets, these processes identify the sensitivity of the climate system to internal variability and its manifestation as variations in the global surface temperature over a range of timescales.

The complexity of the climate system and its interacting energy processes underscore the inadequacy of the IPCC's simple theoretical framework of radiative forcing. The atmosphere and ocean are fluids in motion that interact and whose variability has time constants extending out to a thousand years and more. As a consequence, the IPCC assertions that the climate system has limited internal variability and that increased concentrations of greenhouse gases in the atmosphere will lead to dangerous climate change are simplistic and erroneous.

Radiation Forcing of Climate

Radiation is an important component of the climate system because it is the dominant form of energy exchange between the earth and its surroundings. The sun is the primary source of radiant energy for the earth and solar energy has its maximum intensity at wavelengths that are characteristic of visible light, although the full spectrum does extend from ultraviolet wavelengths to the much longer infrared wavelengths. The earth intercepts solar radiation as a near parallel beam so that the overall intensity is strongest at the surface when the sun appears to be directly overhead. Also, at any point in time, half of the earth's surface is in darkness. The intensity of solar radiation decreases as the angle between the local vertical and the apparent direction of the sun increases. As a consequence, the intensity of solar radiation is greatest over the tropics.

The intensity of the solar radiation varies with the square of the distance from the sun. If the earth's orbit is perfectly circular then the total amount of solar radiation intercepted each day (the global insolation) would not vary through the year. However the orbit is slightly elliptical and the global insolation when the earth is closest to the sun is about 5 percent more than when it is at its farthest point. Overall, the tropics receive more solar insolation than the polar regions but the amount is highly variable with season and location.

Radiation emitted from the earth has its maximum intensity at longer infrared wavelengths. The earth is continually emitting terrestrial radiation to space in all directions and the intensity is largely a function of the emitting temperature. Although we would expect the maximum terrestrial radiation to be emitted from the warm tropics this is not always the case. Where there are deep convective clouds the emission of terrestrial radiation is from the cold cloud tops and is significantly less than neighbouring regions of generally clear skies. Therefore, the emission of terrestrial radiation from the equatorial regions, that regularly have deep convective clouds, is significantly less than the regularly clear skies over the subtropical surface high pressure regions. Although the local emission of longwave radiation to space is much more regular than the input of solar radiation there is still a degree of variability.

The earth's climate is the outcome of a near-balance between absorbed shortwave radiation from the sun and emitted longwave radiation to space. It is emphasised that this is a near-balance because

The variation by latitude and month of the total daily solar radiation (calories/cm²) reaching the top of the atmosphere. The Declination of the Sun marks the latitude where the sun is directly overhead at midday and the seasons correspond to the northern hemisphere (Source: Smithsonian Meteorological Tables, 1966).
(For comparison, 1 cal cm^{-2} day^{-1} is equivalent to 0.48 Wm^{-2} day^{-1}

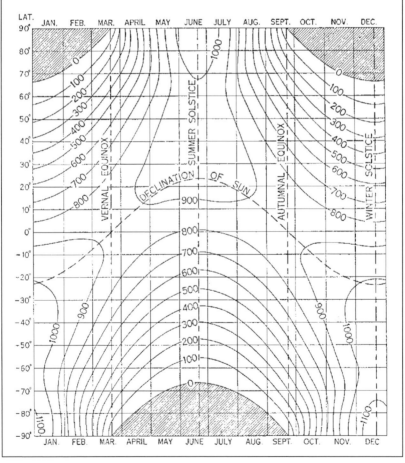

Figure 9:   The seasonal variation of daily insolation at the top of the atmosphere

there is no control to force a balance. The earth's orbital characteristics ensure an ever-changing input of solar energy over the earth's surface and the emissions from the surface, clouds, greenhouse gases and aerosols vary with the characteristics of the location. The processes of the climate system, especially the circulations of the atmosphere and oceans, are constantly working toward achieving a balance, but never succeeding. In addition, interactions between components of the climate system, especially the oceans and the atmosphere, contribute to internal variability that is reflected in local climate fluctuations.

It should also be noted that this emphasis on achieving a near-balance between the absorption of solar radiation and emission of longwave radiation is a defining difference between the arguments being presented here and the radiative forcing hypothesis presented the IPCC. According to the IPCC hypothesis, a stable climate state results from global radiation balance and that any natural or imposed change to the radiation balance will force the climate system to a new climate state that restores global radiation balance. Prior to the commencement of the industrial revolution in 1750 the earth's climate was 'stable' and that there was global net radiation balance at the top of the atmosphere. IPCC asserts, based on the multi-proxy temperature reconstruction of Figure 7, that the climate has been in radiation balance for at least the 750 years leading up to industrialisation. The IPCC has concluded that, as a consequence of anthropogenic increases of greenhouse gas concentrations since industrialisation, the characteristics of terrestrial emissions to space have been changed and rising global temperatures have been a necessary response to maintain global radiation balance.

The simple IPCC hypothesis of anthropogenic radiative forcing loses its potency in a construct where an enhanced greenhouse effect is but one of many variations to which the climate system is continually adjusting as it attempts to restore near-balance of the earth's radiation exchange. The IPCC argument is also diminished in a construct of permanent poleward transport of energy that is essential to restore the imbalance caused by ongoing solar radiation excess in the tropics and deficit over polar regions. It is illuminating to examine the earth's radiation exchange in some detail and identify the seasonally varying geographic radiation imbalances.

Solar radiation

Energy input to the earth's climate system is from solar radiation and is highly variable around the globe. Figure 9 maps the daily solar radiation reaching the top of the atmosphere by latitude and month. The greatest variation is over the polar regions of each hemisphere where the peak values during the summer solstice exceed those of tropical latitudes but there are also many months with little or no solar insolation. It is the seasonal and latitudinal variability of solar insolation at the top of the atmosphere that is the main contributor to the ongoing variability of the climate system.

The latitudinal and monthly variation of solar insolation shown in Figure 9 is a consequence of the tilt (or obliquity) and precession of the earth's axis of rotation and the eccentricity of the earth's orbit around the sun[52]. The alternating six months of darkness over each pole is a consequence of the tilt of the earth's axis of rotation. The current tilt is about 23.5 degrees and total daily darkness at the time of winter solstice in each hemisphere extends to latitude 66.5 degrees. The earth's orbit is slightly eccentric and the earth is currently closest to the sun on 4 January each year. As a consequence, more solar radiation is received by earth during the southern hemisphere summer than during the northern hemisphere summer. The tilt of the earth's axis is slowly precessing and in 11,000 years the earth will be closest to the sun near the northern hemisphere summer solstice.

A little more than 30 percent of the total solar radiation reaching the top of the atmosphere is reflected back to space by clouds, aerosols and the earth's surface. The earth absorbs the remainder of the solar radiation but only about 20 percent is absorbed by the atmosphere and its constituents. The remainder, or 50 percent of the solar radiation intercepted by the earth, is absorbed at the earth's surface.

The local nature of the earth's surface is important for characterisation of how solar radiation is absorbed and how the energy is made available to the climate system. More than 70 percent of the earth's surface is ocean and solar radiation penetrates into and is absorbed by the surface layer to a depth of about 100 metres. Mechanical mixing by winds and wave action ensures that the absorbed energy is mixed through the ocean surface layer. The effective thermal capacity of the ocean surface layer is relatively large and means that the ocean surface temperature responds only slowly to local variations in solar radiation, both from day to day and with seasons.

In contrast to the oceans, solar radiation absorbed by the land surface contributes directly to heating of the surface. Heat will penetrate into the surface but this is only after a temperature gradient is established. As a consequence, the surface temperature over land responds rapidly to solar heating and the solar energy is retained in a layer generally not exceeding a few metres in depth. When compared to the ocean surface the thermal capacity of the land surface is relatively small. Land surfaces heat rapidly under the influence of solar radiation but also cool rapidly when solar radiation is reduced.

---

52 Refer to Figure 6 for an explanation of these terms.

The differing characteristics of land and ocean surfaces are important for understanding how absorbed solar radiation drives the climate system. The relatively small thermal capacity of land surfaces and the temperature response to solar radiation ensures that solar energy absorbed by land surfaces quickly becomes available for transport within the climate system. In contrast, the sea surface temperature responds more slowly as solar energy is absorbed in the ocean surface layers. In this context, the ocean surface layers act as an energy reservoir and are the flywheel of the climate system.

Terrestrial radiation[53]

The greenhouse gases of the atmosphere, together with clouds and aerosols, regulate terrestrial radiation exchange within and from the climate system. As can be seen from the schematic representation in Figure 10, the greenhouse gases, clouds and aerosols of the atmosphere absorb most of the terrestrial radiation emitted upwards by the earth's surface but they themselves emit longwave radiation, both back to earth and to space. Water vapour and carbon dioxide are very effective greenhouse gases and the downward directed terrestrial radiation received at the surface emanates from a low altitude. Unlike carbon dioxide, which is well mixed through the atmosphere, water vapour has its highest concentrations near the surface but is distributed more deeply through the troposphere in regions of ascending air.

The effectiveness of water vapour as a greenhouse gas and emitter of terrestrial radiation can be judged by the differing overnight cooling rates of dry continental and moist maritime locations in the subtropics. When the air is moist the high water vapour concentration is effective in emitting downward directed terrestrial radiation and the surface is slow to cool. However with dry air and clear skies the downward directed terrestrial radiation is reduced and surface cooling is greater. This example demonstrates the importance of water vapour as an emitter and absorber of terrestrial radiation in the boundary layer and low troposphere.

The greenhouse effect is the reduction in upward directed terrestrial radiation between the surface and the top of the atmosphere. From Figure 10 we can readily see that the global greenhouse effect of the earth's atmosphere (including greenhouse gases, clouds and aerosols) is the reduction from 390 W/m^2 emitted at the surface to the

---

53 Terrestrial radiation is the name given to radiation emitted by the earth's surface and greenhouse gases, clouds and aerosols in the atmosphere. Where solar radiation is emitted by the extreme heat of the sun and is most intense in visible wavelengths the maximum intensity of terrestrial radiation (emitted at relatively cool temperatures) is at longer infrared wavelengths. Terrestrial radiation is also referred to as longwave radiation.

235 W/m^2 actually emitted to space. This is a reduction of 155 W/m^2 and is determined by the emissions from greenhouse gases in the high troposphere (mainly carbon dioxide because the air in the high troposphere is generally very dry) and cloud tops that are much colder than the earth's surface.

The IPCC anthropogenic greenhouse warming hypothesis proposes that increased concentrations of carbon dioxide and other well-mixed anthropogenic greenhouse gases will initially cause a reduction in the terrestrial radiation being emitted to space, thus enhancing the greenhouse effect. If there were less terrestrial radiation to space then the climate system would be retaining energy and expected to become warmer with time. When the effective radiating temperature of the earth has warmed sufficiently then the emission of terrestrial radiation to space would return to previous levels, as is required by radiation balance at the top of the atmosphere.

In the first instance it is illustrative to evaluate the ongoing net terrestrial radiation loss of the combined troposphere and boundary layer. We note that of the 390 W/m^2 terrestrial radiation emitted by the

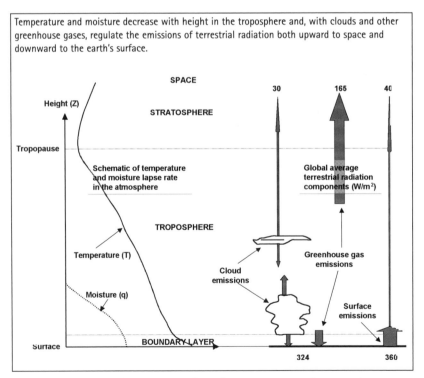

Figure 10: A schematic representation of temperature, moisture and terrestrial radiation in the atmosphere

underlying surface, 350 W/m² are absorbed by the troposphere. However, there are 195 W/m² emitted to space and 324 W/m² emitted back to the surface. This is a net ongoing loss of terrestrial radiation amounting to 169 W/m². Even with an absorption of 67 W/m² of solar radiation (see Figure 1) this is still an ongoing radiation deficit to the troposphere and boundary layer of 102 W/m², a continuous cooling of about 1°C per day. If it were not for other compensating processes the radiation deficit would cause the troposphere to cool.

It is important to note that a reduction in the emission of terrestrial radiation to space will not immediately impact on the temperature of the troposphere. Even if an increase in the concentration of greenhouse gases in the atmosphere causes a reduction in the emission of terrestrial radiation to space then the net radiation loss of the troposphere and boundary layer is only reduced marginally. Calculations based on the global average energy budget suggest that a doubling of the concentration of carbon dioxide in the atmosphere will reduce the net loss to space by about 4 W/m² and increase the downward emission at the surface by about the same amount. Radiative cooling of the troposphere will continue to be at about 1.0°C per day.

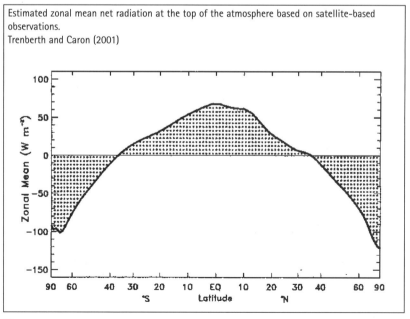

Estimated zonal mean net radiation at the top of the atmosphere based on satellite-based observations.
Trenberth and Caron (2001)

Figure 11: Variation by latitude of annual average net radiation at the top of the atmosphere

Poleward energy transport

The one-dimensional energy budget model used by the IPCC in support of its explanation of radiative forcing ignores the meridional net radiation imbalance at the top of the atmosphere. As previously noted (see Figure 9), at the top of the atmosphere the daily solar insolation varies over the globe with latitude and season. When averaged over the annual cycle there is net inflow of radiation energy to earth over the tropics and net outflow over higher latitudes and polar regions, as shown in Figure 11[54]. This meridional variation in net radiation at the top of the atmosphere clearly points to processes other than radiation being important in determining surface temperatures of the earth. It is the circulations of the oceans and atmosphere that continually transport energy poleward. As a consequence of the continuous poleward transport of energy, tropical surface temperatures are cooler and polar temperatures are warmer than they would be under local radiation balance.

The latitudinal distribution of net radiation at the top of the atmosphere varies seasonally, especially over middle and high latitudes, according to the seasonal variability of solar insolation (see Figure 9). For example, over polar regions there is excess solar insolation over terrestrial radiation about the time of midsummer but maximum net terrestrial radiation to space during the darkness of winter. Thus, for a short period during summer these polar regions receive excess solar insolation despite the annual mean deficit of net radiation at the top of the atmosphere. As a consequence of the overall annual deficit of net radiation over polar latitudes, there is a requirement for an import of energy from the tropics to maintain regional energy balance. However the annual cycle of net radiation at the top of the atmosphere ensures an annual cycle in the rate of meridional energy transport with maximum transport during winter months.

Kevin Trenberth and Julie Caron of the National Center for Atmospheric Research, USA[55] have calculated the annual average rate of meridional transport of energy by the atmosphere and oceans. They used satellite observations of solar and longwave radiation for the period 1985–1989 to provide an estimate of the meridional distribution of the annual average net radiation at the top of the atmosphere (Figure 11). The meridional variation of net radiation at the top of the atmosphere determines the total meridional transport required by the circulations of the atmosphere and oceans to achieve global radiation balance for the earth's climate system. The meridional energy transport

54 Trenberth, K.E. and J.M. Caron, 2001. Estimates of meridional ocean and atmospheric heat transports., J. of Climate, vol 14, 3433–3443.
55 Trenberth, K.E. and J.M. Caron, 2001. Ibid.

by the atmosphere was independently calculated from the meteorological fields of the NCEP/NCAR Reanalysis Dataset[56]. The ocean transport is the residual difference between the overall meridional energy transport required for radiation energy balance at the top of the atmosphere and the energy transport by the atmospheric circulation as calculated using meteorological observations.

The peak poleward transport of energy by the atmospheric circulation was calculated by Trenberth and Caron as about 5.0 PW (5.0x10[15] joules/sec) in both hemispheres but at slightly different latitudes (43°N and 40°S respectively). From these calculations, the atmospheric circulation dominates the overall meridional energy transport and accounts for about 78 percent of the total transport in the northern hemisphere and 92 percent in the southern hemisphere. The atmospheric energy transport has a well-defined annual cycle with maximum transport during winter months of each hemisphere. The annual cycle is significantly more pronounced in the northern hemisphere and this is consistent with the warmer summers and colder winters of that hemisphere. The peak ocean transport is at about 35 degrees latitude in each hemisphere.

The assumption that there is a requirement for a globally averaged annual net radiation balance at the top of the atmosphere provides a simple and seemingly plausible closure for the computation of selected vertical and meridional energy components of the climate system[57]. The magnitudes of the major energy flows within the climate system, particularly the longwave radiation components and the meridional transport by the circulations suggest that, for the purpose of calculating these gross magnitudes, the assumption is justified. For calculation of long-term trends within the climate system, however, the assumption is open to challenge. The complexities of the interactions between radiation and other components of the climate system (including variations in ocean heat storage, polar ice mass, clouds and land surface characteristics) suggests that small errors in calculating energy flows through the climate system will accumulate as escalating biases in local temperature.

---

56 The USA's National Center for Environmental Prediction (NCEP) and National Center for Atmospheric Research (NCAR), with international cooperation have reanalysed all of the available meteorological and associated data back to 1958 with a consistent methodology to produce a high quality set of global meteorological fields for climate research.

57 Trenberth, K.E. and A. Solomon, 1994. The global heat balance: heat transports in the atmosphere and the ocean. Climate Dynamics, vol 10, 107–134. Ibid. They recognise that the global mean net radiation may depart from zero but suggest that the departure should be less than 1 W/m^2. Nevertheless, their data covering 1988 showed a global imbalance of 4.1 W/m^2 and they made adjustments, mainly to the absorbed solar radiation, to achieve global balance.

The magnitude of the energy reservoir of the ocean surface layers and their impacts on the atmospheric circulation are particularly difficult to assess. The El Niño-Southern Oscillation (ENSO) and its variations of sea surface temperature over the tropical Pacific Ocean is just one of the phenomena that is observed to significantly contribute to variability of meridional energy transport, at least on the interannual to decadal timescales. The changing energy storage of the ocean surface layers is difficult to quantify and is yet to be satisfactorily included in computer models.

In summary, the IPCC's anthropogenic greenhouse warming hypothesis is simple and at first sight appealing. However the hypothesis is inadequate because the one-dimensional energy balance model underestimates the complexity of the processes that regulate energy flow through the climate system. The hypothesis is incorrect in that it makes global radiation balance a necessity of the climate system rather than a state about which the climate system is continually fluctuating, depending on the interaction of its internal processes. The IPCC hypothesis ignores the changing thermal capacity of the earth, at least that part of the earth's mass (the land surface and the mixed surface layer of the oceans) that need to be heated to raise the mean temperature of the troposphere. The hypothesis also ignores the important range of processes of the climate system that are affected and could potentially compensate for changed terrestrial radiation, including the meridional transport of energy and cloud cover associated with the weather systems.

The Atmosphere

The atmosphere is a relatively thin layer of gaseous fluid that envelops the earth. At sea level, atmospheric pressure varies but has an average value of about 1013.2 hPa. Surface pressure is higher within anticyclones and lower with cyclones. The density of the atmosphere at sea level is about 1.2 Kg/m³. Both pressure and density decrease with altitude. At 12 km (approximately the cruising altitude of jet aircraft) the pressure is only about 200 hPa and the density is 0.31 Kg/m³. Also, on average, temperature decreases from about 15°C at the surface to about –55°C at 12 km altitude.

The principal chemical constituents of the atmosphere are nitrogen (78.09 percent), oxygen (20.95 percent), argon (0.93 percent) and carbon dioxide (0.03 percent). Water vapour is also a significant component whose concentration varies with temperature and atmospheric pressure. The saturation vapour pressure of water vapour (that is, the maximum concentration before condensation begins) varies at sea level from about 6 hPa near 0°C to about 42 hPa near 30°C. Over tropical oceans it is not unusual to have temperatures near 30°C and humidity near 80 percent so that the actual water vapour concentration is near 0.023 percent. Both carbon dioxide and water vapour are naturally occurring greenhouse gases.

For most climate considerations the important layers of the atmosphere are the boundary layer, the troposphere and the stratosphere (see Figure 10). The boundary layer is next to the earth's surface and may be only a few tens of metres thick under a night-time inversion or several kilometres thick over a hot desert. The characteristic of the boundary layer is that it is well mixed, either by eddies associated with wind or by dry thermals associated with a relatively warm underlying surface. The boundary layer is therefore deeper during strong winds and during afternoon heating but becomes shallower as the wind eases or as the surface temperature cools, such as overnight or under cloudy conditions. Energy transferred to the atmosphere from the underlying surface, both as conduction of heat or evapotranspiration of moisture, is constrained within the boundary layer and not mixed higher into the atmosphere by wind or dry thermals.

The troposphere is the layer of the atmosphere between the earth's surface and the stratosphere and is where weather systems occur. Its depth varies from 14–15 km in the tropics to less than 10 km over polar regions. Ascending motion in weather systems is either within buoyant convective clouds observed over the tropics and over higher latitudes in summertime, or as extensive and deep layer clouds associated with the rising air of mid-latitude cyclones. Ascending air within clouds is saturated and cools at about 6.5°C/km, very close to the normal temperature lapse rate (rate of decrease of temperature with altitude) through the troposphere.

In the cloudless regions of the troposphere the air generally is subsiding to compensate for the ascent that is taking place within clouds. However the unsaturated air warms at 10°C/km as it subsides, which is greater than the 6.5°C/km lapse rate of the troposphere. As a consequence, the compensating subsidence will tend to warm the atmosphere.

The tropopause is the separation between the troposphere and the stratosphere. The stratosphere is a layer of stable air above the troposphere. Its temperature is either constant with height or increases with height. Any vertical motions, whether upward or downward, are inhibited by negative buoyancy forces. The density of air in the stratosphere continues to decrease with altitude.

The circulation of the troposphere is the primary pathway for export of excess energy from the tropics to polar regions. The export varies seasonally as the circulation responds to the varying pole to equator temperature gradient. Also, the annual cycle of energy transport for each hemisphere is not regular because there is significant year to year variability of weather and climate. In order to develop an understanding of the pathways by which the atmosphere transports energy poleward it is necessary to review some of the main processes of the atmosphere and how they interact. The one-dimensional energy balance model, as used by the IPCC in its Third Assessment Report to justify radiative forcing, ignores meridional transport of energy as an important process in the climate system necessary to achieve global radiation balance at the top of the atmosphere.

Mixing of energy through the troposphere

As we have noted previously, the clouds, greenhouse gases (including water vapour and carbon dioxide) and aerosols in the atmosphere regulate the emission of terrestrial radiation from earth. Also, most of the terrestrial radiation emitted by the land and ocean surfaces is absorbed within the troposphere. The atmosphere also emits terrestrial radiation, both upward and downward. The relatively warm temperatures of the lower troposphere govern the downward emissions by the greenhouse gases that reach the earth's land and ocean surfaces. The temperature of the overlying air is generally colder than the earth's surface and there is a net flow of longwave radiation from the surface to the troposphere. (An exception is over the cold polar ice sheets during winter when the air may be 30°C warmer than the surface and the net flow of radiant energy is downward from the atmosphere toward the surface.) The cold temperatures in the high troposphere govern the magnitude of the upward emission to space and the effective radiation temperature of earth is much colder than the measured surface temperature.

The upward and downward longwave radiation emissions from the troposphere are partially offset by absorption of longwave radiation emitted from the underlying surface and by direct absorption of solar radiation. Overall, however, as a consequence of the greenhouse gases, clouds and aerosols, the troposphere emits more radiation than it absorbs and the effect is to cool the troposphere. Based on the global average radiation loss of about 102 W/m^2 (see Figure 1) the average cooling is about 1.0°C per day. The cooling is more in the tropics and less over the colder high latitudes where the concentration of water vapour (the principal atmospheric greenhouse gas) is less.

The radiation cooling of the troposphere is ongoing but there is no direct evidence that such cooling is occurring. The transfer of energy (both heat and latent energy) from the warm underlying surface offsets the radiation loss. Although the radiation loss is distributed throughout the troposphere the replacement energy from the underlying surface accumulates in the atmospheric boundary layer, especially over the tropical oceans. The processes that distribute heat and latent energy from the boundary layer are deep tropical convection and the atmospheric overturning of the Hadley Cells. These processes regulate the temperature distribution through the troposphere. The essential roles of these processes further underscore the inadequacy of the simple radiative forcing model used by the IPCC, which infers that temperatures are an outcome of the local radiation balance.

The IPCC, in its one-dimensional energy balance model of the climate system, indicates that the excess energy of the earth's surface is mixed through the troposphere by 'thermals and evapotranspiration'. Such an explanation is completely unsatisfactory because the small scale eddies that transfer heat and latent energy from the oceans and land surfaces to the atmosphere cannot penetrate beyond the atmospheric boundary layer, where the energy tends to accumulate. The explanation fails to describe how the boundary layer energy is further distributed through the troposphere. The distribution of energy through the troposphere is integral to the energy flow through the climate system.

Saturated air, ascending buoyantly in the protected updraughts of deep tropical convective clouds, will conserve moist static energy as it rises through the troposphere. The updraught air spreads out at the tropopause where moist static energy and dry static energy are equivalent. During ascent, heat and latent energy are converted to potential energy as the air cools and vapour condenses. Throughout the tropics there is compensating subsidence and potential energy is being converted to heat. The rate of conversion of potential energy to heat balances cooling through radiation loss.

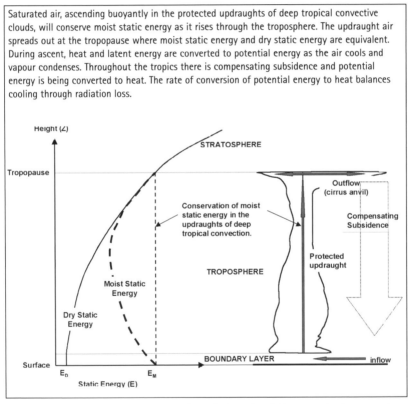

Figure 12: A schematic representation of the role of deep tropical convection in distributing boundary layer energy through the troposphere

58   The dry static energy relates to unsaturated air and is the sum of the enthalpy ($c_p{}^*T$) and potential energy (g^*Z), ie, $E_D = c_p{}^*T + g^*Z$; where E_D is dry static energy, c_p is specific heat at constant pressure, T is temperature, g is the force of gravity and Z is altitude. It is closely linked to the potential temperature of the air. During vertical motion of unsaturated air its energy is conserved and there is an exchange between enthalpy and potential energy. Rising air gains potential energy but cools while subsiding air loses potential energy but warms.

As shown schematically in Figure 12, the dry static energy[58] of the troposphere increases with height. Upward mixing of energy by thermals and mechanical mixing is confined to the atmospheric boundary layer and, when well mixed, will only achieve constant dry static energy in that layer. Within the troposphere, above the boundary layer, mechanical mixing by turbulence will actually mix dry static energy down the gradient towards the surface. Turbulence will tend to cool the upper layers of the troposphere. Therefore, contrary to the claim by IPCC, thermals and evapotranspiration are an inadequate description of how heat and latent energy accumulating in the boundary layer are distributed through the troposphere.

The mechanism for distributing energy from the tropical boundary layer through the troposphere to offset radiation cooling and provide export to higher latitudes was identified by Professors Herbert Riehl and Joanne Malkus of the University of Chicago, USA, in 1958[59]. In their explanation they described the role and mechanisms of deep tropical convection for transforming heat and latent energy of the boundary layer to potential energy in the high troposphere. They demonstrated the crucial role of buoyant ascent of mass within the protected updraughts of deep tropical convection. Further, the Hadley Cells of the tropics represent the meridional overturning of the atmosphere that is necessary to complete the cycle of energy distribution. The essential meridional circulation further underscores the inadequacy of the IPCC one-dimensional energy balance model for the climate system.

The basics of the Riehl and Malkus explanation are contained in Figure 12. The moist static energy[60] of the tropical boundary layer is significantly higher than the corresponding dry static energy because of the presence of the latent energy of water vapour. Cumulus clouds are formed when unsaturated air in the boundary layer is lifted to saturation. However mixing with the unsaturated air surrounding the cloud results in evaporation of cloud material, cooling and loss of buoyancy. As a general rule, the cumulus clouds of the tropics have a short life cycle and limited penetration into the troposphere. Cumulus clouds mix energy and moisture to the level of minimum moist static energy but no further.

Riehl and Malkus proposed that the distribution of energy to the middle and high troposphere could be achieved if the updraughts of the deep tropical convection are protected from the unsaturated

---

59 Riehl, H. and J.S. Malkus, 1958. On the heat balance of the equatorial trough zone. Geophysica, vol 6, Nos. 3–4 pp503–538.
60 The moist static energy is the sum of the dry static energy (E_D) and latent energy (L^*q), ie, $E_M = c_p^*T + L^*q + g^*Z$; where L is latent heat of vaporisation and q is the specific humidity of water vapour.

environment. Air in the boundary layer with moist static energy, E_M, if protected from the environment will conserve its energy as it rises buoyantly to the tropopause. During the buoyant ascent within the protected 'hot towers', heat and latent energy are converted to potential energy and the ascending air is cooled at the moist adiabatic lapse rate[61]. The precipitation rate represents the vigour of the mass flow through the convective cloud.

The ascending updraught air of deep convection loses buoyancy at the tropopause[62] and spreads out from the cloud top as the frequently observed cirrus anvils. The air at the tropopause is so cold that even at saturation it holds negligible water vapour. The arriving air has the same temperature (heat) and potential energy as the environment and neither have significant latent energy. The air arriving at the troposphere displaces air already at the level and there is a compensating subsidence in the surrounding air.

Protected updraughts can only occur if they are surrounded by saturated cloud mass that prevents mixing with the unsaturated environment air. If mixing does occur then some of the condensate evaporates and causes the periphery of the cloud to cool and lose buoyancy. It is for this reason that deep tropical convection generally only occurs in organised mesoscale weather systems, tropical storms and tropical cyclones. Mixing with unsaturated environmental air erodes isolated convection clouds and buoyant ascent to the tropopause is not achieved.

Away from deep convective clouds the air is subsiding to compensate for the ascent that is taking place within the clouds. However, the subsiding air is unsaturated and potential energy is converted to heat. If the troposphere were not cooling we would anticipate that the subsiding air would also conserve its energy and eventually arrive back at the surface with a dry static energy E_D equivalent to E_M at which it started its buoyant ascent. However, because the subsiding air is continually subject to radiation cooling, it is losing energy and eventually arrives back at the boundary layer in the subtropics with dry static energy less than E_D and moist static energy less than E_M. As the air returns

---

61 The moist adiabatic lapse rate corresponds to the temperature decrease with height as saturated air is lifted vertically without exchanging energy with its surroundings (ie, it is 'protected'). Latent energy is being released because condensation is taking place and the rate of cooling is less than for the equivalent unsaturated air. Overall, because the air is cooling and losing water vapour (condensation falls to the ground as precipitation), it is losing heat and latent energy and gaining potential energy. In the lower troposphere the moist adiabatic lapse rate is about $-6.5°C/km$, very similar to that of the surrounding cloud-free air.

62 The tropopause marks the lower boundary between the stratosphere (where temperature is constant or increases with height) and the troposphere.

equatorward, to the focus of deep tropical convection, it again accumulates heat and latent energy from the underlying surface.

The Hadley Cells represent a large-scale tropical overturning of the tropical troposphere that facilitates the transfer of heat and latent energy from the boundary layer through the troposphere. Broadly, there are three components to the Hadley Cell circulation:

(i) Boundary layer air flows towards the equator in both hemispheres and is warmed and acquires moisture, especially if it is passing over an ocean surface. The accumulated energy (the sum of the heat and latent energy) is confined to the layer beneath the Trade Wind temperature inversion until the air reaches the equatorial regions favoured for deep tropical convection.

(ii) Ascending air is concentrated in a relatively few deep convection clouds of the intertropical convergence zone, tropical storms and active monsoon regions. Heat and latent energy are transformed to potential energy within the protected updraughts of deep convection as air buoyantly ascends to the high troposphere.

(iii) In the high troposphere, air flowing poleward from the deep convection, transports the potential energy to the subtropics and gently subsides. Potential energy is converted to heat in the subsiding air over the subtropics. Part of this heat is used locally to offset radiation cooling and part is available for export to middle and higher latitudes.

In their calculations of the equatorial energy budget, Riehl and Malkus noted that the direct conversion of heat and latent energy within the protected updraughts was not sufficient to meet the requirements for observed energy export from equatorial latitudes. They suggested that there was additional overturning within the deep convective clouds as a consequence of organised downdraughts within the clouds. These downdraughts occur because cloud air mixing with the surrounding dry air at the cloud boundary is cooled by evaporation of cloud material. The cooling is sufficient to provide negative buoyancy to cloud air and, with evaporation of further cloud material, it will sink to the boundary layer. The downdraughts tend to cool and dry the boundary layer air in the cloud wake and, as a consequence, enhance local exchange of heat and evaporation from the underlying surface. In addition, because of the need for mass balance within the convection system, the actual mass flow and energy distribution by the updraughts is greater within the convection clouds than is suggested by calculations based solely on the broadscale Hadley Cell circulation.

Convective overturning is a very efficient process for distributing energy from the atmospheric boundary layer to the troposphere. However the process is regulated by the need for buoyancy of the air within the protected updraughts. The warming of the troposphere as a consequence of the widespread compensating subsidence can only raise the temperature to that of the updraught air because at this temperature buoyancy ceases and the overturning ceases. Very clearly, the convective overturning and rate of transfer of energy from the boundary layer to the troposphere is regulated by the rate of radiation loss from the troposphere. However the rate of radiation loss does not determine the temperature of the troposphere because the temperature within the protected updraught is an upper limit for the environment.

The processes for mixing heat and latent energy through the troposphere identifies a major error in the IPCC hypothesis for anthropogenic global warming. The IPCC radiative forcing model assumes that only terrestrial radiation processes will be affected by changing greenhouse gas concentrations. Consequently, a reduction in emission of terrestrial radiation to space will be reflected as a warming of the troposphere. But the need to maintain buoyancy within the deep convective updraughts clearly demonstrates the fallacy of such an assumption. Any reduction in radiative cooling of the troposphere will reduce the convective overturning and the rate of transfer of energy between the underlying surface, the boundary layer and the tropical troposphere. That is, energy is retained in the underlying surface and opens up for consideration the need to take account of the enormous thermal capacity of the ocean surface layer and the land surface.

The thermal capacity of the oceans will significantly slow the impact of changing net terrestrial radiation at the earth's surface caused by changing concentrations of anthropogenic greenhouse gases. Before there is any change to the temperature of the troposphere there is a need to change the temperature of the underlying surface. Moreover, the need for the underlying surface to warm prior to any warming of the troposphere exposes the fallacy of 'positive feedback effects' due to water vapour and clouds. Water vapour concentrations and cloud distributions are an outcome of the temperature, energy exchange processes and mass circulation of the troposphere. Very definitely, the earth's terrestrial radiation emissions are an outcome of the temperature and lapse rate of the troposphere and do not determine them.

Temperature trends of the tropical troposphere

One of the controversies of the anthropogenic global warming debate is the apparently different temperature trends observed over recent decades between the earth's 'surface' and that of the low to middle troposphere. The assessments of recent global warming (see for example Figure 3) have been derived using a combination of data from meteorological instruments over land that measure temperature near the ground and from ships, buoys and satellites that estimate surface temperature over the oceans. For the surface temperature data there is a marked warming since the mid-1970s of about 0.15°C/decade. By contrast, estimates of the temperature trend of the low to middle troposphere using satellite instruments suggest little (0.05°C/decade) temperature change since 1979 when such data became available[63].

The apparent difference in temperature trend between the surface and the low to middle troposphere is an embarrassment to IPCC and the anthropogenic global warming proponents because it is not readily explained. Indeed, one of the mechanisms by which the climate system is expected to amplify the anthropogenic greenhouse effect is through 'water vapour and cloud feedbacks'. Under this scenario the warming troposphere is expected to hold a higher concentration of water vapour (a natural greenhouse gas) and thereby enhance the anthropogenic warming effect. However, the evidence for tropospheric warming cannot be demonstrated, at least over the past two decades of surface warming.

The IPCC has identified several potential sources of error in the low to middle troposphere temperature estimates obtained from satellites. Notwithstanding these, and the more recent derivations applying additional corrections, there remains a close correspondence between the satellite-derived temperature trends and the trends derived from the archive of weather balloon data for the same period. IPCC acknowledges that there are probably real differences in the temperature trends but that they have no satisfactory explanation for the differences[64].

In looking for an explanation as to why the surface and the low to middle troposphere are changing temperature at different rates it is necessary to look at physical processes that are either coupling the temperature responses or causing them to respond differently. There are two important reasons why the temperature of the troposphere might be decoupled from that of the surface. Firstly, dry static energy increases with height and energy is inhibited from mixing upwards from the surface even though the surface is at a higher temperature

63 Christy, J.R., R.W. Spencer, W.B. Norris, W.D. Braswell and D.E Parker, 2003. Error estimates for version 5.0 of MSU-AMSO bulk atmospheric temperatures. J. of Atmos and Ocean Tech. vol 20 (May) pp613–629.

64 IPCC TAR, 2001. Climate Change 2001: The Scientific Basis. (Eds, J.T. Houghton, Y. Ding, D.J. Griggs, M. Noguer, P.J. van der Linden, X. Dai, K Maskell and C.A. Johnston) Cambridge University Press, pp123 and 729.

than the troposphere. Secondly, due to its greenhouse gases, the troposphere is essentially an energy sink because of the ongoing net longwave radiation loss at all latitudes. Riehl and Malkus identified deep tropical convection as the process that is distributing heat and latent energy from the atmospheric boundary layer to the troposphere in order to compensate for the net radiation loss.

At least over the tropics, deep convection regulates the temperature of the troposphere. Buoyant updraughts in the protected 'hot towers' will only occur while the ascending air is warmer than the clear air surrounding the clouds. The temperature of the updraught is determined by the moist static energy of air being drawn in from the boundary layer, while the temperature of the air surrounding the deep convective clouds is regulated by the rate of compensating subsidence of the surrounding air. A steady state of tropical convective over-turning will be established when the rate of conversion of heat and latent energy within the updraughts balances the sum of the net radiative cooling within the tropics and the rate of meridional transport of energy out of the tropics towards polar regions. The temperature of the troposphere cannot exceed the updraught temperature.

When the moist static energy of the boundary layer increases (for example by abnormally warm sea surface temperature, or significantly lower surface atmospheric pressure (as in tropical storms), then the buoyancy of the deep tropical convection will increase (ie, slightly warmer updraught temperature). The increased buoyancy leads to more vigorous overturning of the tropical atmosphere. More vigorous overturning, however, will cause more net heating in the subsiding air, leading to atmospheric warming and a loss of convective buoyancy. A new steady state will become established with the temperature of the troposphere slightly higher than it was previously. Conversely, if the warmest sea surface temperatures over the ocean cool then there will also be a cooling of the temperature of the troposphere.

Empirical studies show that, for the oceans, deep tropical convection only occurs over the warmest tropical waters[65]. Significant deepening of the convection does not occur until mean sea surface temperatures exceed 27°C and the deepest convective clouds occur when the sea surface temperature is greater than 28°C. Monthly sea surface temperatures above 30°C are rarely observed over the ocean and there are strong theoretical reasons to suggest a practical upper limit of sustained sea surface temperature. If deep tropical convection only occurs when the sea surface temperature is between 28°C and 30°C then this is a strong control over the temperature of the tropical troposphere.

65   Ramanathan, V. and W. Collins, 1991. Thermodynamic regulation of ocean warming by cirrus clouds deduced from observations of the 1987 El Niño. Nature, vol 351 pp27–32.

One of the reasons suggesting a practical upper limit to sea surface temperature is that the saturated vapour pressure of water increases exponentially with temperature. Latent energy exchange to the atmosphere by evaporation dominates the surface energy exchange as the sea surface temperature approaches 30ºC. This means that when sea surface temperatures are typical of those in the tropics there is a very rapid increase in evaporation with any increase in sea surface temperature. As a consequence of the rapid increase in latent energy loss (and the high thermal capacity of the ocean surface layer) the warmest tropical sea surface temperatures respond only slowly to changing net radiation at the surface. Therefore, the predominant impact of any decrease in net terrestrial radiation loss at the surface (an outcome of increased atmospheric greenhouse gas concentration) will be to increase latent energy exchange with the atmosphere. Only a small fraction of the additional radiant energy will be available to heat the ocean surface mixed layer.

Empirical data also suggest that increased energy export from the tropics by the ocean and atmospheric circulations and, possibly, increased cloud albedo are strong feedback effects that also act to regulate equatorial sea surface temperatures[66]. Overall, the deepest 'hot towers' of tropical convection occur where the boundary layer air has maximum energy. This is mostly over the equatorial oceans with the warmest sea surface temperature, such as the equatorial western Pacific Ocean and the intertropical convergence zones of the Pacific, Atlantic and Indian Oceans. It is these warm waters, and the deep convection that they support, which maintain the temperature structure of the tropical troposphere.

It should not be a surprise, therefore that satellite and balloon observations of low and mid-tropospheric temperatures, at least over the tropics, do not show a warming trend over recent decades. The updraught temperatures of the protected 'hot towers' are confined to a narrow range linked to the warmest sea surface temperatures. Through updraught buoyancy considerations, temperatures of the tropical troposphere are also constrained within a narrow range. The overturning tropical troposphere is characterised by relatively few very vigorous 'hot towers' but each is transporting large quantities of mass and energy from the boundary layer to the upper levels of the troposphere.

The tropical troposphere is a region of remarkably uniform temperature as the subsiding air, originating with the characteristics of the air flowing out from the upper layers of the deep convective clouds, maintains the tropospheric temperature against radiative cooling. This is in marked contrast to the air of the boundary layer where the air

66 Sun, D-Z and K.E. Trenberth, 1998. Coordinated heat removal from the equatorial Pacific during the 1986–87 El Niño. Geoph. Res Lett., vol 25, No.14 pp2659–2662.

takes on the temperatures and horizontal temperature gradients of the underlying land and ocean surfaces, accumulating energy until it achieves buoyancy and is organised within deep tropical convection.

There is a very wide spatial variation of surface temperature over the tropics related to the net surface radiation and the properties of the underlying land (including vegetation and evapotranspiration) and ocean. Temperatures typically increase from the cooler middle latitudes to the sub-tropics and equatorial regions. In some parts there are also significant zonal variations such as the typical 6°C variation from east to west across the equatorial Pacific Ocean. In contrast, at any level of the lower, middle and high troposphere the temperature variation in any direction within the tropics and into the subtropics is less than about 2°C. This latter pattern of uniformity is remarkable for an area that covers more than one half of the globe, but perfectly understandable in the context of overturning related to convection and radiation.

Clearly, the different trends of the surface temperature and the tropospheric temperature are identifying a real thermal separation of the boundary layer from the overlying troposphere. The basis of the separation is the increase of dry static energy with altitude in the troposphere that prevents vertical mixing by eddies. The physical separation can be readily identified as temperature inversions in the low troposphere that separates the boundary layer and its thermal mixing from the gently subsiding air of the troposphere. The temperature inversion is present across sub-tropical anticyclones and at the height of cumulus cloud tops that characterise tropical Trade Winds.

The satellite data do show slight increases in middle tropospheric temperature during El Niño events, when there is a more extensive area of the equatorial Pacific Ocean with sea surface temperatures above 27°C[67]. There is also evidence that during these events more energy is exported to middle and high latitudes by the atmospheric circulation[68].

Surface energy exchanges

The simple one-dimensional energy balance model used by IPCC to justify its radiative forcing hypothesis (see Figure 1) is unrealistic in its portrayal of processes at the earth-atmosphere interface. The IPCC model suggests that the heat and latent energy exchange between the underlying surface and the atmosphere is a direct response to the imbalance of solar and terrestrial radiation at the surface. Such a proposal is at odds with the physics of the surface energy exchange processes. In particular, the IPCC model takes no account of the

67   Spencer, R.W. and J.R. Christy, 1990. Precise monitoring of global temperature
     trends from satellites. Science, vol 247 pp1558–1562.
68   Trenberth, K.E., D.P. Stepaniak and J.M. Caron, 2002. Interannual variations
     in the atmospheric heat budget. J.Geoph. Res. vol 107, No. D8 pp4(1–15).

effective thermal capacity of the underlying surface that is an important component for regulating surface temperature.

Over land, solar radiation is absorbed at the surface (either by vegetation or soil) and heat penetrates into the underlying material by conduction. The effective thermal layer for land surfaces is shallow and temperature variations are confined to within a metre of the surface. The change in thermal capacity of the underlying land surface is small when considering long-term surface temperature changes and only introduces a small error to the calculations. A further consideration is changes to vegetation or soil moisture that will also have an effect on surface temperature through evapotranspiration and latent energy loss to the atmosphere.

Oceans make up about 70 percent of the earth's surface. When considering the earth's energy exchange processes it is erroneous to neglect the changing energy storage of the oceans' surface layer. Solar radiation penetrates the ocean surface and is gradually absorbed with depth. The ocean surface layer, to a depth of about 100 metres, acts as an energy reservoir as it absorbs solar radiation and the energy is mixed within the layer by wave action. Energy is lost from the ocean surface layer by exchange across the sea-air interface and through transport by the ocean currents. In some regions of the oceans it is upwelling of cold waters from depth that also strongly modifies surface temperature. The depth of the mixed layer and its effective thermal capacity (taking account of transport and upwelling) are factors that determine the rate at which surface temperature responds to changes in net surface radiation. The ocean is a fluid and the effective thermal capacity of the surface layer will vary with wind speed (the mixing depth) and upwelling.

To demonstrate the importance of ocean upwelling in regulating sea surface temperature we need only look at the equatorial Pacific Ocean where the sea surface temperature difference between the west and the east is about 6°C. This is despite solar radiation at the top of the atmosphere being the same in both regions. An important factor that keeps sea surface temperatures lower in the east is the upwelling of cold water along the South American coast and extending westward along the equator. In addition, energy is transported towards the western equatorial Pacific Ocean by the westward directed ocean surface current. The temperature of the ocean surface (the sea surface temperature[69]) is therefore a function of the local solar insolation,

---

69 Sea surface temperature, or SST, is the name generally given for the ocean temperature at the sea-air interface. Direct observations are made of the engine cooling water intake below the surface level whereas satellite observations are of the 'skin' (or radiating surface) temperature. The 'skin' temperature can be warmer than the underlying temperature in sunlight but cooler at night due to ongoing evaporative cooling.

internal ocean processes of the surface layer (the effective thermal capacity, including the mixing depth and vertical and horizontal heat transports) and the rate of loss across the ocean-atmosphere interface.

The predominant energy loss from the oceans is by way of heat and latent energy exchange at the ocean-atmosphere interface. Winds blowing across the ocean surface promote evaporation (latent energy loss) and conduction of heat to the overlying atmospheric boundary layer. There is also net longwave radiation loss from the surface. Greenhouse gases, clouds and aerosols in the lower troposphere, absorb most of the emissions of longwave radiation from the surface but some of the emission also passes through the atmospheric window[70] and is lost to space. The emission of longwave radiation from the surface is mostly offset by absorption of downward directed longwave radiation from above. The warm temperatures and greenhouse gases in the lower layers of the troposphere, particularly water vapour, contribute to the relatively high intensity of back-radiation at the surface. The net longwave radiation loss at the surface is relatively small when there is a deep humid boundary layer.

The primary regulator of heat and latent energy exchange between the ocean surface and the atmospheric boundary layer is sea surface temperature. Evaporation (ie, latent heat exchange) increases very rapidly with temperature because of the exponential increase of saturated vapour pressure with temperature[71]. Over tropical oceans, where sea surface temperature approaches 30ºC, it is latent heat exchange through evaporation that is the dominant form of energy loss from the ocean surface. Under most situations the ocean surface is warmer than the overlying atmosphere and there is transfer of energy from the ocean to the atmosphere.

Over equatorial latitudes the daily solar insolation varies over a relatively narrow range through the year. Tropical sea surface temperatures also vary over a small range although in some locations, such as the eastern equatorial Pacific Ocean, there is significant seasonal variation contributed by local upwelling of cold water and other internal dynamic processes of the ocean surface layer. Over middle and high latitudes the sea surface temperature varies strongly with the seasonal cycle of solar insolation. During winter the energy exchange with the overlying atmospheric boundary layer exceeds incoming solar radiation and the surface layer of the ocean cools; in

---

70 The 'atmospheric window' is the range of wavelengths over which radiation emitted by the earth's surface passes through the atmosphere and its greenhouse gases without absorption. For all other wavelengths the emitted radiation is absorbed by the greenhouse gases of the atmosphere

71 The Clausius-Clapeyron relationship between saturation vapour pressure and temperature.

summer it is the incoming solar radiation that dominates the energy balance and the surface layer of the ocean warms.

There are two significant differences between land-to-air and ocean-to-air energy exchanges. Firstly, as noted above, solar radiation at the land surface only penetrates into the soil by conduction and the absorbed solar energy is confined mainly to a shallow surface layer. As a consequence, land surfaces warm more than ocean surfaces at corresponding latitudes because, for the latter, the energy is mixed through a deeper ocean layer. Secondly, evapotranspiration from plant and soil surfaces is less effective for evaporative cooling than is evaporation from open water surfaces. Nevertheless, empirical data suggests that when there is high rainfall to support vegetation and evapotranspiration there is an effective upper limit to daily maximum air temperature of about 34°C[72]. By comparison, sea surface temperature rarely exceeds 30°C.

The effect of evapotranspiration on surface and near-surface air temperature is important in the discussion of the extent to which land-use change has affected climate. We have previously noted the urban effect on surface air temperature as asphalt and concrete replace tracts of trees and grasslands. The need to feed an expanded global population has meant that very large land areas have been converted from forests and woodlands to managed grasslands. Grasses extract soil moisture from a shallow layer compared to the deep roots of trees. Therefore, where previously the trees were drawing subsoil moisture for ongoing evapotranspiration there is now less evapotranspiration and a marked seasonal cycle corresponding to the growing season. The extent to which these land-use changes have affected global temperature records and the recent apparent warming remains controversial but the direction of any change is unambiguous. Any reduction in evaporation over time resulting from land-use changes, such as over areas of deforestation, will be observed as a warming trend at the surface.

The moist static energy of the boundary layer determines the temperature of the rising air in the updraught of deep convective clouds. When warm moist air rises buoyantly from the atmospheric boundary layer in deep convective clouds the updraught temperature has to be warmer than the temperature of the troposphere. There is very little spatial variation in the tropospheric temperature over the tropics. As a consequence, deep convection occurs over those regions of the tropics with maximum moist static energy of the atmospheric boundary layer. That is, the deep convection occurs over those regions that provide the maximum buoyancy for air ascending in the protected updraughts.

72 Priestley, C.H.B., 1966. The limitation of temperature by evaporation in hot climates. Agric. Meteorol., 3, pp 241–246

Deep convection over tropical oceans generally occurs when the sea surface temperature exceeds 28°C[73]. Here, evaporation from the underlying surface and the warm temperature maintains high values of moist static energy in the boundary layer. Over land, although tropical and subtropical surface temperatures are generally much hotter than the oceans, the surface evapotranspiration and boundary layer latent energy are generally less than over the warm tropical oceans. Despite the high temperatures, the moist static energy of the boundary layer is generally too low to support vigorous deep convection over tropical lands.

An exception is when the land surface becomes very wet and additional heat and moisture from the surface add to the moist static energy of air moving from adjacent tropical oceans. Such a situation arises within the monsoon circulations of the tropics. The monsoon circulation is sustained so long as the boundary layer moist static energy over the land is sufficiently high to maintain the buoyancy

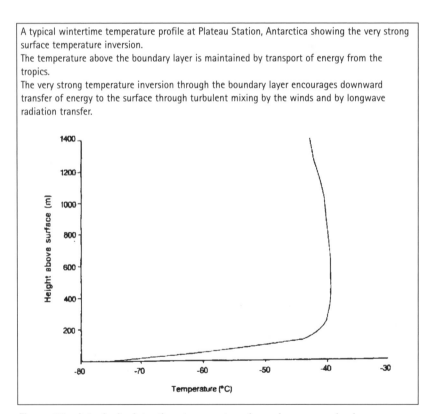

A typical wintertime temperature profile at Plateau Station, Antarctica showing the very strong surface temperature inversion.
The temperature above the boundary layer is maintained by transport of energy from the tropics.
The very strong temperature inversion through the boundary layer encourages downward transfer of energy to the surface through turbulent mixing by the winds and by longwave radiation transfer.

Figure 13: A typical wintertime temperature inversion over polar ice

73   Ramanathan, V. and W. Collins, 1991, Ibid.

necessary for deep convection. Under these conditions the deep convection is located preferentially over tropical land.

Polar ice and surface energy exchange

Surface energy exchanges over polar regions, especially over the extensive snow and ice surfaces of winter are very different to those over the tropics. During the months of the polar winter there is little or no incoming solar radiation. A characteristic of the boundary layer air is the development of a very strong temperature inversion as shown in Figure 13. Instead of the expected decrease of temperature with height there is a very strong increase of temperature with height in the lowest few hundred metres. This occurs because the atmospheric circulation that transports energy from the tropics tends to maintain the temperature of the troposphere against radiative cooling. Air over the poles is very dry and, as a consequence, the downward emission of longwave radiation from the atmosphere to the surface is relatively small. The emission of longwave radiation from the surface continues to be relatively strong and there is a cooling of the surface. In the darkness of winter, the surface cools until there is a balance between its own longwave emission, the heat conduction from below, and downward longwave radiation emitted from the warmer but dry overlying troposphere.

The contrast between wintertime and summertime surface air temperatures over the Arctic shown in Figure 14 is very dramatic.

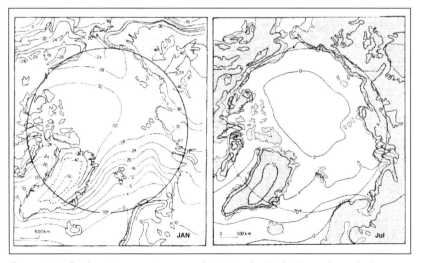

Figure 14: Surface temperature over Arctic and sub-Arctic regions during winter and summer

During winter, surface temperatures are well below 0ºC over the frozen Arctic Ocean and the adjacent sub-Arctic land area. Temperatures are coldest over the high Greenland Plateau but extremely cold temperatures are also experienced over Siberia and the higher parts of Alaska. Mildest temperatures are experienced to the west and north of the Scandinavian Peninsula where the relatively warm North Atlantic Current penetrates into the Arctic Ocean. However in summer it is only the high Greenland Plateau and the central Arctic Ocean near the North Pole that remain below freezing. Elsewhere, particularly over the sub-Arctic land areas that are experiencing long day length and significant daily insolation, the surface temperatures are above freezing.

The accumulation of snow over the Arctic and Antarctic surfaces is primarily an outcome of the atmospheric circulation. As with tropical convection, vertical motion of saturated air causes cooling, condensation and an increase in potential energy. In the cold polar regions, the latent energy that is released during the sublimation of water vapour to ice crystals is taken up as potential energy by the ascending air. The ice crystals fall and accumulate on solid surfaces as snow. The snow remains in place until it melts, evaporates or is transported to the coast within a glacier. The accumulations of snow over the permanent ice sheets of Antarctica and Greenland, which are permanently colder than 0ºC and not subject to summer melting, are strongly related to the annual precipitation rate. Over other regions where there is summer melting, including mountain glaciers, the factors governing the balance between expansion and contraction of snow and ice mass are more complex. Annual snow fall contributes to expansion of the ice mass but absorption of solar radiation and conduction of heat from the atmosphere provide energy for ice-melt when temperatures are above 0ºC.

The formation of sea-ice over the Arctic Ocean and the coastal margins of the Antarctic continent are also an outcome of surface energy exchange processes. During the darkness of winter there is net heat loss from the ocean surface to the atmosphere, particularly from net longwave radiation loss but also from evaporation and eddy heat exchange. These processes work to cool the ocean surface. There is some compensation because winds blowing over the surface tend to mix heat upwards from within the ocean surface layer, but this upward mixing by the water motion is reduced when the surface freezes. The upper surface continues to lose heat to the atmosphere when sea-ice forms and cools to sub-zero temperatures. Heat is transferred through the ice layer by conduction and the sea-ice thickens as water at the interface loses heat and freezes.

The potential for melting of snow and ice over polar regions is limited to those areas where surface temperatures during summertime exceed 0ºC. For the Arctic and adjacent regions (Figure 14 – right-hand panel) we can readily see that over Greenland it is only around the coastal margins that there is potential for seasonal melting. Similarly, direct melting of the upper surface of the Arctic Ocean sea-ice is limited near the North Pole. Where sub-zero temperatures are experienced into summer, ice loss can only occur by direct ablation to water vapour from the snow pack or, in the case of sea-ice, to melting at the ocean interface. Evaporation of sea ice requires approximately 8 times more energy then for melting. Snow melt and ice ablation accelerate when temperatures reach above 0ºC in summer.

The amount of incoming solar radiation actually absorbed by the snow and ice surface depends on the albedo (or reflectance) of the surface. Snow and ice surfaces are generally highly reflective and typically absorb less than 30 percent of incoming solar radiation. However, by contrast, snow-covered forests typically absorb 80 to 90 percent of the solar radiation reaching the surface and this energy greatly assists snowmelt. As a consequence of their high albedo, the ice sheet of Greenland and the sea-ice of the Arctic Ocean persist whereas the snow cover of the boreal forests is perennial. We also note that the coastal sea-ice of the Arctic Ocean bordering the North American and Eurasian continents is fragile. The coastal sea-ice is subject to melting in summer (temperatures reach above 0ºC) and, once exposed, the lower reflectance would ensure that surface waters would absorb additional solar energy and warm. The break up of areas of Arctic sea-ice in summer is assisted by cracking (leads and polynias) that increases the absorption of solar radiation, and by heat transport of the atmospheric and ocean circulations.

In summary, energy exchange at the earth's surface is far more complex, and important to the climate system, than is implied by the IPCC one-dimensional energy balance model. Solar radiation absorbed by the ocean surface layer is not immediately available to the atmosphere and the ocean surface layer is more accurately described as an energy reservoir whose capacity can vary with time. Changes in the strength of the ocean circulation and surface upwelling are factors that contribute to the determination of sea surface temperature patterns and the rate of exchange of energy between the oceans and atmosphere. Surface energy exchanges over polar regions vary significantly with the seasons and, particularly in summer, are dependent on the nature of the surface, whether open ocean, sea ice or land. Exposed land surfaces, especially boreal forests, absorb significantly more solar energy into the earth's climate system than does reflective snow. Similarly, open ocean absorbs more solar radiation than does sea-ice cover.

Seasonal patterns of tropical convection

The annual cycle of solar heating is related to the tilt of the earth's axis of spin and the earth's orbit around the sun and so there is a degree of regularity to tropical rainfall. People living within the tropics and subtropics depend on the reliability of seasonal rains for fresh water and food production. Notwithstanding the overall control by the orbital characteristics, there is significant variability in the annual patterns of tropical and subtropical rainfall. Characteristic patterns of drought and flood recur with enormous human impact. Famine and disease continue to plague many communities on a regular basis despite major efforts related to planning, early warning and international relief measures that are being adopted to mitigate impacts. Rainfall variability associated with El Niño events is recognised as the cause of characteristic regional drought over some areas and flooding over others. These events persist on timescales of seasons to more than a year.

Droughts lasting a decade or more are also characteristic of the subtropics and middle latitudes of all continents. Decadal length droughts of modern times include the 1930s Dust Bowl event of North America, the prolonged Sahel drought of Africa during the 1960s and early 1970s, and the persisting drought conditions of eastern Australia from 1895 through 1903. As with the shorter-lived El Niño events, there is mounting evidence that these long duration climate anomalies are also linked to persisting patterns of sea surface temperature that change regional circulations of the atmosphere and rainfall patterns.

There is a propensity for deep tropical convection to occur in regions that provide maximum buoyancy for the 'hot towers'. Temperature is relatively uniform throughout the tropical troposphere and so it is boundary layer energy variations that determine potential buoyancy of the convective updraughts. In the tropics, maximum buoyancy occurs where boundary layer energy is a maximum and there is a close relationship between regions of maximum boundary layer energy and fields of deep convection. The moist static energy of the air is made up of three components: heat, latent energy of water vapour and potential energy. In the atmospheric boundary layer, where potential energy is small, energy is highest where there is high humidity to maximise the combined effects of temperature and atmospheric water vapour.

Over the tropical oceans there is abundant water vapour available from evaporation and, as a consequence, maximum energy occurs over the warmest waters. A map of annual mean sea surface temperature that identifies the regions of warm water is at Figure 15. It is no surprise that the warm waters of the equatorial western Pacific Ocean and the equatorial seas within the Indonesian Archipelago are sources of abundant moisture and are also regions of high rainfall. Elsewhere over the tropical oceans the easterly winds of each hemisphere converge in a narrow band near the equator and it is here that we find boundary layer air with high energy. This intertropical convergence zone, as it is called, is linked to the larger scale atmospheric circulation and moves with the seasons.

The availability of local moisture to provide energy for the boundary layer is more limited is more limited over tropical land areas. Evapotranspiration from tropical forests returns a proportion of previous rainfall to the boundary layer air but there is an overall reliance on transport of moisture from surrounding oceans. Unlike the oceans, where solar radiation is absorbed through a deep surface layer, land surface temperatures respond more directly to solar heating. Solar radiation absorbed over land surfaces is very quickly transferred to the overlying boundary layer, but more as heat than as latent energy (the ratio depending on the surface wetness and vegetation cover) and consequently humidity is low relative to the oceans.

During summer months, near-surface air over the tropical land areas is generally hotter than the oceans at the same latitude but the lower humidity means that the energy of boundary layer air over the land is often lower than that of the oceans. As a counterbalance to the lack of boundary layer moisture, the hotter air over land causes surface air pressure to fall slightly. Thus, on surface pressure charts we observe 'heat lows' over the hot subtropical landmasses during summer. The warmer air associated with the 'heat low' is only as deep as the boundary layer thickness but the pressure drop is sufficient to

Climatological values of sea surface temperature for the tropics and middle latitudes.

The warmest waters are over the equatorial western Pacific and eastern Indian Oceans. A tongue of cold water resulting from wind-induced upwelling extends westward along the equator from the South American coast.

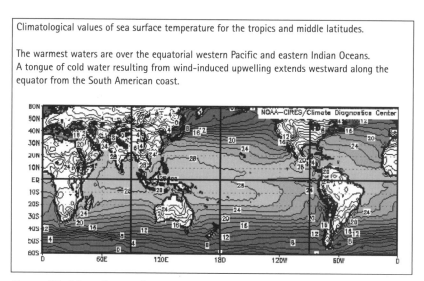

Figure 15: Map of annual average sea surface temperature

cause an inflow of air and moisture from the surrounding oceans. Over Africa and South America, continents that straddle the equator, the zone of maximum solar heating moves north and south with the seasons and localities receive maximum rainfall during seasonal maximum solar heating.

For Asia and Australia the subtropical landmasses are offset from the equator and convection over each responds quite differently to the seasonal cycle of heating. The Australian landform is relatively flat and boundary layer air and with high moisture is drawn from the surrounding oceans by 'heat lows' that form over the hot land. Very vigorous individual convective clouds are often observed, especially in the late afternoon and evening, but the lifecycle of clouds is short and rainfall from isolated showers. This is because the troposphere is very dry. Consequently, as convection penetrates into the dry air the cloud material mixing at the edges is quickly evaporated, drawing heat from the air that is being mixed in from the cloud surroundings. The air above the boundary layer is moistened by the convection but, without significant precipitation, there is very little net release of latent energy to support deep atmospheric overturning. The air drawn to the summertime continental 'heat low' of Australia is confined to a shallow depth below 1,500 metres[74]. Overall, the limited moisture inflow and the high summer temperatures result in Australia being the driest inhabited continent.

By contrast, subtropical parts of the Asian continent have striking topographical landforms, including much mountainous terrain. In summer the high mountain surfaces are heated by solar radiation and become warmer than surrounding air at the same level. The warmer air on the mountain slopes is relatively buoyant and establishes airflow up the slopes that reinforces the continent-wide boundary layer inflow of air to the 'heat low'. In contrast to the situation over Australia, a deep inflow of boundary layer air is established that provides adequate moisture to sustain deep convection.

Satellite instruments have enabled precise mapping of the locations and seasonal movement of deep tropical convection. The tops of deep convective clouds are very cold and the magnitude of the longwave radiation emitted from cloud tops to space (outgoing longwave radiation – OLR) is relatively low. In contrast, the outgoing longwave radiation emission from regions with few deep convective clouds is much higher because the emission originates from greenhouse gases lower in the troposphere that are relatively warm. The distribution of deep convective clouds can be observed from patterns of outgoing longwave radiation to space as measured by

74   Troup, A.J., 1974. The mean flow at the boundary of the Australian continent. Aust. Meteor. Mag., vol 22, No. 3 pp6–66.

satellites. Tropical areas that have seasonally low outgoing radiation have frequent deep tropical convective clouds whereas areas those with high values of average outgoing longwave radiation are relatively cloud-free.

Seasonal maps of outgoing longwave radiation have been produced[75], for example as shown in Figure 16. These clearly identify

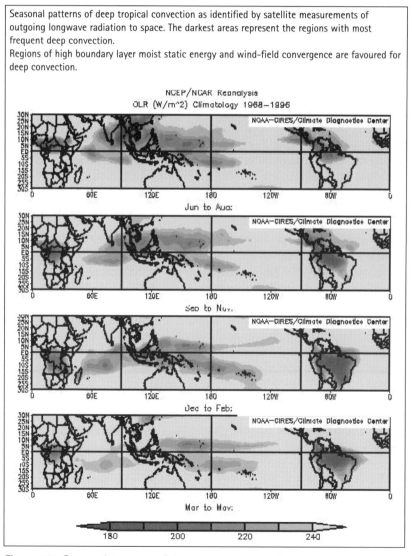

Seasonal patterns of deep tropical convection as identified by satellite measurements of outgoing longwave radiation to space. The darkest areas represent the regions with most frequent deep convection.
Regions of high boundary layer moist static energy and wind-field convergence are favoured for deep convection.

Figure 16: Seasonal patterns of deep tropical convection

75 Meehl, G.A., 1987. The annual cycle and interannual variability in the tropical Pacific and Indian Ocean regions. Mo. Wea. Rev., vol 115, pp27–50.

the north-south peregrinations of deep convection over equatorial Africa and South America. It is of interest that for Africa the southward extension of convection during the austral summer and autumn (December through May) is greater than the northward extension during the boreal summer and autumn (June through November). Over the Americas, however, deep convection extends northward along the cordillera of the Central America isthmus during the summer-autumn period.

Over the Asia-Pacific region there is a distinctive eastward march to the seasonal pattern of tropical convection. During boreal summer, deep convection extends over a broad region from southern China to the Malay Peninsula and southwestward across eastern and southern India. The region of deep convection contracts from the west and north as the focus shifts southeastward over the Malay Peninsula and the Philippines during boreal autumn. By austral summer the deep convection has shifted further southward and eastward, extending from Indonesia through New Guinea and southeastward into the South West Pacific. The deep convection persists over the South West Pacific during the austral autumn. There is then a reversion of the focus to South East Asia as the Asian monsoon is re-established during the boreal summer.

In addition to the south and eastward march of the focus of tropical convection across the Asia Pacific region there are regions of the Indian and Pacific Oceans where deep convection persists through the year. One is in the central Indian Ocean south of the equator. Here, the sea surface temperature remains relatively warm throughout the year[76] and the seasonal winds associated with the Asian monsoon enhance the convergence of warm moist boundary layer air.

Deep convection clouds are also present throughout the year over the warm surface waters of the equatorial western Pacific Ocean. The intensity of the deep convection fluctuates seasonally north and south of the equator. North of the equator deep convection is most active during the boreal summer, and extends far eastward to the central Pacific along the intertropical convergence zone at about 10°N. South of the equator deep convection is most active during the austral summer and extends southeastward from New Guinea along the South Pacific Convergence Zone, a regional band of convergence within the southeasterly winds of the boundary layer that flow over the western Pacific Ocean.

76 Harrison, D.E. and N.K. Larkin, 1996. The COADS sea level pressure signal: a near global El Niño composite and time series results, 1946–1993. J. of Climate, vol 9, pp3025–3055.

The climatological seasonal patterns of deep convection are readily disrupted by changes to seasonal patterns of sea surface temperature. For example, during an El Niño event warm ocean waters extend further eastward to the central and eastern equatorial Pacific Ocean. The focus of deep convection is dragged eastward with the warm waters and away from the land areas of South East Asia and the islands of Indonesia. In contrast, during a La Niña event, when the warm tropical waters of the equatorial Pacific Ocean contract to the western margins, deep convection is focussed over land areas of South East Asia and even extends southward across northern and eastern Australia.

The tropical wind systems are known as monsoons, or steady winds, and this implies a degree of regular seasonal recurrence driven by the annual cycle of solar heating. Notwithstanding this steadiness, slight changes in tropical sea surface temperatures can significantly change the patterns of seasonal moisture convergence in the boundary layer, and hence patterns of deep convection. There is abundant evidence of changing patterns of deep convection on the interannual timescale linked to El Niño and La Niña events. There is also evidence that patterns of tropical sea surface temperature vary on decadal and longer timescales and these have been sufficiently pronounced to significantly alter relative populations of local fish species[77] as well as the overlying atmospheric circulation. Attempts have been made, with a degree of success, to link decadal variability of sea surface temperature patterns with changing precipitation patterns[78] and to ascribe regional drought to a persisting anomaly of tropical sea surface temperature[79] using computer models. There is evidence that anomalous tropical sea surface temperatures that persisted during the 1930s caused the devastating 'dust bowl' drought of North America, and that interactions between the atmosphere and the land surface increased its severity[80].

Clearly, sea surface temperature patterns have a major control over the location and intensity of seasonal deep convection over the tropics and subtropics. As a consequence, variability of sea surface temperature patterns on interannual to decadal timescales also affects

77 Chavez, P.C., J. Ryan, S.E. Lluch-Cota and M. Niquen C., 2003. From anchovies to sardines and back: multidecadal change in the Pacific Ocean. Science, vol 299, pp217–221.
78 Meehl, G.A., J.M. Arblaster and W.G. Strand Jr, 1998. Global scale decadal climate variability. Geophys. Res. Lett., vol 25, No. 21, pp3983–3986.
79 Hoerling, M. and A. Kumar, 2003. The perfect ocean for drought. Science, vol 299, pp691–694.
80 Schubert, S.D., M.J. Suarez, P.J. Pegion, R.D. Koster, and J.T. Bacmeister, 2004. On the cause of the 1930s dust bowl. Science, vol 303, pp1855–1859.

precipitation distribution and intensity, especially over tropical land areas. Any long-term changes in tropical patterns of sea surface temperature will not only affect the availability of energy to the atmosphere for export from the tropics but it will also affect the distribution of tropical precipitation.

Transport of Energy by the Atmosphere

Local net radiation at the top of the atmosphere varies from the equator to the poles, with excess solar input over the tropics and deficits over polar regions (see Figure 11). The circulations of the atmosphere and oceans work to transport the excess energy received over the tropics to higher latitudes. Also, in both hemispheres most of the transport is by the atmosphere (approximately 78 and 92 percent in the northern and southern hemispheres respectively). We will now review the processes that regulate the rate of poleward transport of energy by the atmosphere. These processes are linked to the earth's rotation and seasonal variations of the pole to equator temperature gradients of each hemisphere.

There are three primary forms in which energy is transported from the tropics to polar regions: as potential energy, as heat, and as latent energy (water vapour). Potential energy is transported whenever there is meridional overturning of the troposphere because potential energy increases with altitude. In the high troposphere, air with high potential energy is transported in the direction of the airflow but in the low troposphere less potential energy is being returned in the compensating airflow. The overturning circulation produces no mass transfer but there is a net transport of potential energy in the direction of the high level flow.

Potential energy transport in the tropics is by way of the overturning Hadley Cells of each hemisphere. Air in the tropical boundary layer extracts heat and latent energy from the underlying surfaces and becomes warmer and more humid as it moves equatorward. At the intertropical convergence zone, and other regions favoured for deep convection, heat and latent energy are converted to potential energy in the ascending branch of the Hadley Cells (the buoyant updraughts of deep convection). As compensation for the average equatorward flow of air at low levels there is high level return flow towards the poles. Overall, there is a transport of heat and latent energy towards the equator and a transport of potential energy toward the poles.

Across the tropics, at any altitude above the boundary layer, there is very little horizontal temperature variation. The tropical troposphere is barotropic and the poleward transport of heat by the horizontal winds is inconsequential. Although there is considerable variability within the horizontal wind-flow over the tropics, heat transported poleward over one region will be compensated by returning wind with similar temperature, and heat content, over another region.

There are also significant quantities of latent energy exported from the tropics because of the strong gradients of water vapour (specific humidity) across the tropics, especially in the lower troposphere. The troposphere is very humid near the intertropical convergence zone and other regions of deep convection because water vapour and cloud material are lifted vertically in the deep convective clouds and mix at the cloud margins with environmental air. Smaller cumulus clouds are also able to mix water vapour vertically through the lower troposphere. By contrast, over the subtropics the subsiding air of the Hadley Cells brings dry air downward from high altitudes. Air moving towards the subtropics transports water vapour poleward and air moving towards the equator returns dry air because of the moist air over the equatorial regions and dry air over the subtropics. Overall, there is a net transport of water vapour (ie, latent energy) poleward from the tropics.

Away from the tropics, over the earth's middle and higher latitudes, there are prevailing westerly winds and strong horizontal temperature gradients. The troposphere of the middle latitudes is baroclinic. Here, direct transport of heat by the horizontal winds is the dominant mechanism for poleward transport of energy. Air that moves towards the poles is warm and air that moves towards the equator returns cooler air. Also, in contrast to the tropics, the westerly winds are relatively strong. Planetary waves, called Rossby Waves, are a feature of the atmospheric circulation and are characterised as great meanders in the horizontal westerly wind flow. There are generally between three and five Rossby Waves around the hemisphere and the successive poleward and equatorward components of the flow support the poleward transport of energy. Maps of geopotential heights (metres) of the 500 hPa pressure surface for the winters of 1997–1998 (an El Niño year) and 1998–1999 (a La Niña year) at Figure 17 show the difference in Rossby Wave patterns under the different tropical forcing regimes.

In contrast to the tropics, potential energy is actually transported equatorward over middle latitudes. This comes about because of the weather systems (cyclones and anticyclones) that are typical of middle latitudes. To a large extent the westerly jet streams of the high troposphere balance the strong horizontal temperature gradients brought about by cold air over polar regions and warmer air over the subtropics. The mathematical details of the thermodynamics are complex but broadly relate meridional temperature gradients with vertical wind shear (the increase of horizontal wind speed with height). The balance requires an increase in vertical wind shear when the horizontal temperature gradient strengthens, and vice versa. However a pronounced equatorward excursion of cold polar air sets up a density difference along the direction of the westerly airflow and this upsets the thermodynamic balance. When the thermodynamic balance is

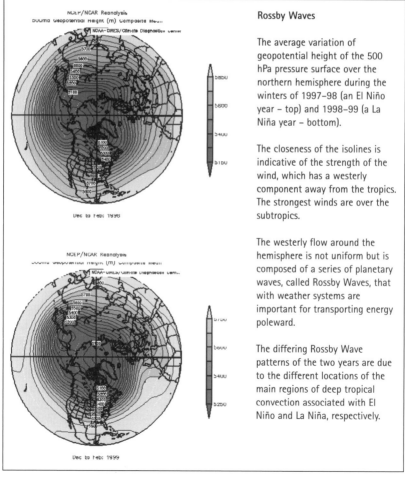

Rossby Waves

The average variation of geopotential height of the 500 hPa pressure surface over the northern hemisphere during the winters of 1997–98 (an El Niño year – top) and 1998–99 (a La Niña year – bottom).

The closeness of the isolines is indicative of the strength of the wind, which has a westerly component away from the tropics. The strongest winds are over the subtropics.

The westerly flow around the hemisphere is not uniform but is composed of a series of planetary waves, called Rossby Waves, that with weather systems are important for transporting energy poleward.

The differing Rossby Wave patterns of the two years are due to the different locations of the main regions of deep tropical convection associated with El Niño and La Niña, respectively.

Figure 17: Planetary waves (Rossby Waves) of the atmospheric circulation

upset the cold air begins to sink and downwind the warm air begins to rise.

Cyclones that are a typical feature of daily weather charts are manifestations of the movement of cold polar air towards the equator. The sinking cold air has its origins in the high troposphere and it converts potential energy to heat as it moves equatorward. The rising warm air has its origins in the warm moist boundary layer and as it rises it cools to saturation. Where clouds develop in the rising air, either as buoyant convective clouds or extensive layer clouds, heat and latent energy are converted to potential energy as the air moves poleward. The average overturning has high level air moving equatorward and sinking and low level air moving poleward and rising such that there is a net equatorward transport of potential energy, or indirect circulation. Notwithstanding the reverse overturning, the magnitude of transport of heat and latent energy by cyclones is greater than the equatorward transport of potential energy and there is a net poleward transport of energy.

Near the poles the thermodynamic balance between wind shear and temperature gradient is no longer relevant. The horizontal wind pattern is similar to a vortex centred on the pole and horizontal temperature gradients are weak. Subsidence, and conversion of potential energy to heat, is the major process for offsetting radiation loss within the polar troposphere, particularly during the darkness of winter. On average there is a direct overturning circulation over polar regions that is strongest in winter. During summer the polar vortex often breaks down and cyclonic weather systems are active over polar regions. Heat and latent energy are imported from the middle latitudes by the horizontal wind circulation associated with these storms. Local overturning within the storms makes the heat and latent energy available to offset net radiation loss within the troposphere over the polar regions.

Overall, excess solar radiation absorbed in the tropics, especially in the ocean surface layer, follows a very complex path before being radiated to space from the polar troposphere. Different factors regulate the rate of flow at each stage of the journey. Over the tropics surface temperature regulates the net emission of longwave radiation from the surface, and surface temperature also regulates the rate of conduction of heat and evapotranspiration from the earth's surface to the boundary layer. The need for buoyancy of the updraughts in deep convection limits the distribution of heat and latent energy from the tropical boundary layer to the tropical troposphere. Over the middle and high latitudes it is the characteristics of the Rossby Waves and the frequency and intensity of weather systems that regulate the transport of energy from the tropics to polar regions. However the strength of the westerly wind flow of the middle latitudes is an outcome of the strength of the overturning circulation of the tropical Hadley Cells. In order to understand this linkage it is necessary to consider the role of conservation of atmospheric angular momentum in airflow and the exchange of angular momentum between the atmosphere and the earth.

Atmospheric angular momentum

The primary zonal wind fields of the troposphere are the easterly winds of the tropics and the westerly winds of middle and higher latitudes, as shown in Figure 18. There are variations to this overall pattern due to local geographical and seasonal influences but the regions of easterly and westerly winds are a characteristic feature of both hemispheres and all seasons. The characteristic pattern is an outcome of both the overturning circulations necessary to distribute heat and latent energy from the tropical boundary layer and the Rossby Waves and cyclonic weather systems necessary to transport excess energy from the tropics to polar regions. The earth's rotation has a major control over the pattern of zonal winds, including the middle latitude westerly winds and embedded Rossby Waves, because of the concept of angular momentum that is linked to the rate of spin around the north-south axis of rotation.

Each unit of mass of the earth has angular momentum related to the product of its tangential speed and the radial distance to the axis of rotation. On the surface of the earth the tangential speed decreases from the equator to the poles because, although the rate of rotation is constant everywhere, the radial distance from the surface to the axis of rotation decreases. The angular momentum of the earth's surface also decreases with distance from the equator because it is the product of the radial distance from the axis of rotation and the tangential speed, each of which decrease with distance from the equator. However, the atmosphere is a fluid that envelops the solid earth and is not constrained by solid body rotation. If the wind has a westerly component it is spinning faster than the earth below but if it has an easterly component it is spinning slower than the earth below.

When we consider airflow moving away from or toward the pole it is necessary to take account of angular momentum, including the exchange of angular momentum between the atmosphere and the underlying surface of the earth due to frictional stresses. A parcel of air at the equator is at the maximum distance from the axis of rotation. If it is moving with the same rotation rate as the earth it has no zonal wind speed relative to the earth. Away from the equator a similar air parcel that has no zonal wind relative to the underlying surface has less angular momentum than the air parcel at the equator because of the shorter radial distance to the axis of rotation. Air in the atmospheric boundary layer tends to take on the same zonal wind speed as the underlying surface because of the action of frictional stresses between the earth and the boundary layer air.

Above the boundary layer, where friction with the underlying land or ocean surface is not important, a parcel of air that moves away from the equator conserves its angular momentum but the distance to the axis of rotation is decreased. As the air parcel in the troposphere moves further away from the equator towards higher latitudes it increasingly has angular momentum greater than the underlying surface. As a consequence, air that moves poleward increases its rate of spin relative to the underlying surface. That is, as air moves away from the equator the westerly wind speed relative to the underlying surface increases. Through similar arguments, air in the troposphere

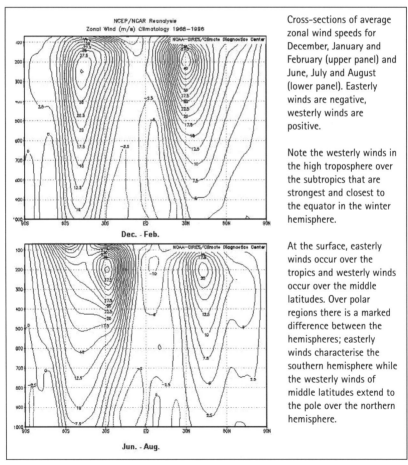

Cross-sections of average zonal wind speeds for December, January and February (upper panel) and June, July and August (lower panel). Easterly winds are negative, westerly winds are positive.

Note the westerly winds in the high troposphere over the subtropics that are strongest and closest to the equator in the winter hemisphere.

At the surface, easterly winds occur over the tropics and westerly winds occur over the middle latitudes. Over polar regions there is a marked difference between the hemispheres; easterly winds characterise the southern hemisphere while the westerly winds of middle latitudes extend to the pole over the northern hemisphere.

Figure 18: Cross-sections of average zonal wind in the atmosphere

that moves toward the equator tends to lose westerly speed relative to the earth below and, depending on its initial zonal speed, may become an easterly wind.

A simple calculation demonstrating how the subtropical westerly wind jetstreams of the high troposphere are generated through conservation of angular momentum is at Figure 19. We will follow an air parcel in the high troposphere that is initially over the equator and has no westerly wind speed. As it moves to latitude 30 degrees while conserving angular momentum there is a significant increase in zonal wind speed. At the equator the air parcel has no zonal wind component compared to the underlying earth surface but it does have a tangential velocity of about 460 m/s because of its rotation with the earth. If the air parcel moves to latitude 30 degrees while conserving angular momentum its tangential speed increases to 530 m/s because of the reduced distance to the axis of rotation. However the tangential speed of the underlying surface is only 400 m/s because of the constant rotation rate of the earth and the reduced distance of the surface to the axis of rotation. The speed of the air parcel relative to the underlying surface is now about 130 m/s, and is consistent with the upper limits of zonal wind speeds observed in jetstreams.

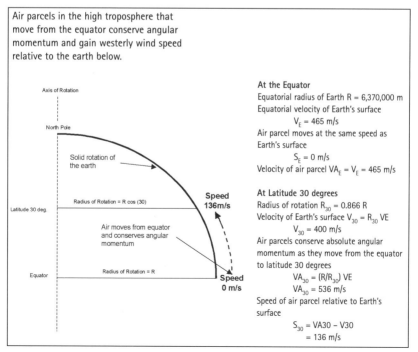

Figure 19: Formation of subtropical jetstreams of the troposphere

Within the atmospheric boundary layer it is necessary to take account of the effects of frictional stress acting between the air and underlying surface. Whenever wind blows over a surface there are frictional stresses between the airflow and the underlying surface. The effect of the frictional stresses on the airflow is to reduce its speed. We are particularly concerned about the cases where the surface airflow has a zonal component, either westerly or easterly. Where the surface airflow has a westerly component it is spinning faster (ie, has greater angular momentum) than the underlying surface. In this situation friction will act to slow the westerly wind and angular momentum will be lost from the westerly winds to the earth. In contrast, a surface airflow that has an easterly component is spinning slower relative to the earth's surface below. Friction will act to reduce the relative speed of the airflow by transferring angular momentum from the surface, thus causing the airflow to spin faster.

Easterly winds are common over the tropics, especially the 'Trade Winds' typical of the tropical oceans. The easterly winds occur because air flowing towards the equator from the subtropics tends to conserve angular momentum and increase the easterly wind speed component relative to the surface. However, the air takes up angular momentum (increasing the rate of spin, or westerly wind component) from the underlying surface because of frictional stresses. As a consequence, within the boundary layer the easterly component of wind speed is constrained as compared to air moving toward the equator above the boundary layer.

The overturning Hadley Cells of the tropical troposphere cause the airflow in the tropical boundary layer to be generally towards the equator and winds take on a relatively weak easterly component. Near the equator, as air rises buoyantly in 'hot towers' within deep convective clouds, it maintains its absolute angular momentum and continues to spin approximately in concert with the underlying surface.

In the high troposphere the air spreads poleward from the 'hot towers' and conserves its angular momentum. As the high troposphere air moves poleward it takes on a westerly component of wind speed relative to the underlying surface. In the high troposphere there is relatively little frictional stress to reduce the westerly wind component and at about 30 degrees latitude in each hemisphere the westerly winds are very strong. The strong westerly winds are the subtropical jet streams observed in the high troposphere. The characteristics of the zonal wind field, especially the tropical surface easterly winds and the strong westerly winds in the high troposphere over the subtropics, are evident in the cross-section of zonal wind speed at Figure 18.

The overturning Hadley Cells of the tropics are a source of relative westerly angular momentum for the high troposphere over middle latitudes. Frictional stresses at the surface act on air moving towards the equator and these cause a transfer of angular momentum from the earth to the atmosphere. The stronger the overturning in the Hadley Cells the more angular momentum that is transferred to the atmosphere. Angular momentum is conserved and distributed through the tropical troposphere by cumulus and deep convective clouds. Angular momentum is also conserved by air that has risen buoyantly in the deep convective clouds and spreads poleward in the high troposphere. However, as the air moves towards the poles the reducing distance from the earth's axis of rotation causes the air to spin up relative to the underlying surface and it acquires relative angular momentum (ie, westerly wind speed) compared to the earth below.

Away from the tropics the Rossby Waves and weather systems transport angular momentum towards the pole and downward to the surface. The poleward transport is achieved because Rossby Waves are not symmetrical and there is a correlation between stronger westerly winds in the poleward airflow and weaker westerly winds in the returning flow toward the equator. Also, the stronger winds ahead of cyclonic storms support poleward transport of angular momentum and the overturning circulation transports westerly momentum downwards to the boundary layer.

The deep westerly winds from the surface to the high troposphere over middle latitudes are a manifestation of transport of angular momentum away from the subtropical jetstreams by the Rossby Waves and weather systems. Friction (wind stress) transfers angular momentum from the atmosphere to earth across the broad band of surface westerly winds extending over middle latitudes.

The surface airflow of the tropical Hadley Cells is towards the equator and the associated easterly winds are continually transferring angular momentum between the earth and the atmosphere. Away from the tropics, angular momentum is returned to the earth by frictional stresses associated with the surface westerly winds. As the rate of overturning of the Hadley Cells increases then the rate of transfer of angular momentum from the earth to the atmosphere also increases. In addition, the average angular momentum of the atmosphere (and the relative speed of the westerly winds) also increases and the earth's rate of rotation slows. The magnitude of tropospheric relative angular momentum and length of day are indicators of overturning and poleward energy transport by the atmospheric circulation.

Angular momentum is being continually exchanged between the earth and the atmosphere because of wind stress in the boundary layer. The atmosphere takes up angular momentum from the earth in regions of tropical surface easterly winds and angular momentum is extracted from the atmosphere over regions of middle and high latitude westerly winds. Pressure forces acting on topographic features (known as mountain torques) also exchange angular momentum between the earth and the atmosphere. The overall exchange of angular momentum between the earth and atmosphere is not constant[81]. At times of increased energy cycling, such as during El Niño events, the angular momentum of the atmosphere increases at the expense of the earth and the earth's rotation slows perceptibly.

Variation of atmospheric energy transport

There are connections between the pole-to-equator energy transport and the relative angular momentum of the troposphere. Firstly, because angular momentum is being continuously transferred between the earth and the atmosphere in the region of tropical surface easterlies, the relative angular momentum of the atmosphere increases with rate of overturning of the tropical Hadley Cells. The westerly winds that are generated by the overturning circulation are in the high troposphere over the subtropics and middle latitudes. The strengthened westerly winds also increase the strength of the vertical wind shear over these latitudes.

Secondly, there is a tendency for the atmosphere to generate a thermodynamic balance between the vertical wind shear and the horizontal temperature gradient. From this connection we note that the strength of overturning of the tropical Hadley Cells is one of the factors regulating the horizontal temperature gradient over middle latitudes. The other factor is the pattern of net radiation at the top of the atmosphere. Excess solar radiation and convective overturning combine to provide energy to warm the tropics; a radiation deficit tends to cool the polar troposphere. The global pattern of net radiation, through its impact on the pole-to-equator temperature gradient, also works to strengthen the vertical wind shear in the troposphere of the middle latitudes.

The thermodynamic balance is relevant to poleward energy transport because there is a higher probability of instability developing within a Rossby Wave when the horizontal temperature gradient and vertical wind sheer of the middle latitude westerly winds strengthen. Such instability (also known as baroclinic instability) favours the development of a regional-scale cyclonic weather circulation. Cyclonic weather systems are more efficient than Rossby Waves for transporting heat, latent energy and angular momentum towards the poles. Middle latitude cyclones are therefore more frequent and more intense during winter when the pole-to-equator temperature gradient is strongest.

81   Hide, R. and J.O. Dickey, 1991. Earth's variable rotation. Science, vol 253, pp629–637.

The relationship between atmospheric overturning and angular momentum generation is observed in the seasonal cycle. The earth's surface and troposphere both cool during winter over middle and high latitudes when the net radiation deficit is at a maximum. The high latitude cooling causes the pole to equator temperature gradient to increase and leads to an enhancement of the poleward transport of energy. There is also a significant increase in the atmospheric angular momentum of the winter hemisphere. The range of the annual cycle of temperature over the northern hemisphere is greater than that of the southern hemisphere and this is reflected as differences in the annual range of atmospheric angular momentum variation of each hemisphere. This difference leads to the annual cycle of global atmospheric angular momentum that can be observed in the upper panel of Figure 20.

There is enhancement of the eddy transfer of heat and moisture from the tropical ocean to the atmospheric boundary layer during an El Niño event, when there is a more extensive area with high sea surface temperatures. El Niño events are also times of increased export of energy from the tropics to middle and high latitudes[82]. Reconstruction of atmospheric angular momentum anomalies over the period 1870–1990, using a computer model of the climate system forced by observed sea surface temperatures, suggests significant multi-decadal variability as a consequence of varying frequency and intensity of El Niño events[83].

There was a notable increase in interannual variability of reconstructed atmospheric angular momentum around 1970 (supported by calculations based on observed atmospheric winds) but this high level may not have been significantly greater than an earlier period from 1900 to 1920. There is some uncertainty about the veracity of sea surface temperature data for the tropics prior to 1945. Changes in sea surface temperature of only a few tenths of a degree Celsius significantly alter the atmospheric angular momentum reconstruction. But clearly, periods of enhanced El Niño activity, as occurred at the turn and into the first few decades of the 20th century and again in the later two decades of that century (see Figure 20, lower panel for the latter), are also periods of enhanced atmospheric angular momentum and poleward energy transport.

The annual cycle of climate is driven by the changing pattern of solar radiation received at the top of the atmosphere. The seasonal

82   Trenberth, K.E., D.P. Stepaniak, and J.M. Caron, 2002. Ibid.
83   Rosen, R.D. and D.A. Salstein, 2000. Multidecadal signals in the interannual variability of atmospheric angular momentum. Climate Dynamics, vol 16, pp693–700.

variation is an important consideration because of the changing latitudes of maximum solar energy input and its impact on local excess or deficit of net radiation. The most dramatic seasonal change in daily solar insolation received at the top of the atmosphere is in the polar regions. Here, no direct solar radiation is received during the darkness of winter months. However at midsummer the peak daily insolation at the top of the atmosphere exceeds that over the equator at the same time. By comparison, the seasonal change in longwave

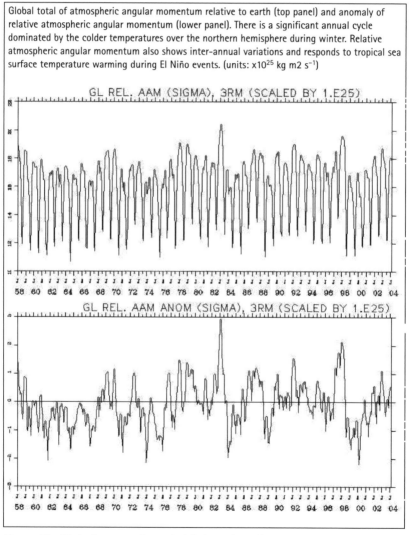

Global total of atmospheric angular momentum relative to earth (top panel) and anomaly of relative atmospheric angular momentum (lower panel). There is a significant annual cycle dominated by the colder temperatures over the northern hemisphere during winter. Relative atmospheric angular momentum also shows inter-annual variations and responds to tropical sea surface temperature warming during El Niño events. (units: x10^{25} kg m2 s^{-1})

Figure 20: Variation over time of global totals and anomalies of relative atmospheric angular momentum

radiation loss to space varies much less. It is the seasonal variation of solar radiation that predominantly determines the local top of the atmosphere radiation imbalance. Over polar regions, therefore, there is an annual vacillation between the enormous radiation deficit during winter to small surplus in summer.

The annual cycle of radiation forcing over the polar regions is large and out of phase between the hemispheres. Over the tropics there is always excess solar radiation and the amplitude of the annual cycle of net top of the atmosphere radiation is muted in comparison to the polar regions. Overall, the seasonally varying and out of phase radiation imbalance represents a complex forcing of the climate system. This forcing is further complicated by the different land-sea distributions of the hemispheres. The amplitude between summer heating and winter cooling of the earth's surface over the northern hemisphere is greater than that of the southern hemisphere because of the greater land area of the northern hemisphere.

Polar surface temperatures get colder over higher latitudes as the radiation deficit increases towards midwinter. The resulting increase in pole to equator temperature difference works toward increasing meridional energy transport. Similarly, approaching midsummer, as solar radiation exceeds longwave radiation loss, temperatures over high latitudes begin to rise. The poleward energy transport abates as the pole-to-equator temperature difference reduces. Over this cycle, the atmospheric circulation, and the ability to transport energy towards the poles, are responding to the seasonal excess and deficits of net radiation over middle and high latitudes.

The seasonal pulse of the atmospheric circulation is transmitted through the climate system as seasonal low-level winds that exert varying stresses on the oceans and drive surface currents. The seasonally varying winds drive the oceans' gyres and the patterns of upwelling and downwelling that modify local temperatures in the surface layers of the oceans. The oceans have much more inertia than the atmosphere[84]. Wind-driven currents at the oceans' surface, although not deep, persist after the winds have eased. Of course, the oceans themselves respond to the seasonally varying imbalance of solar radiation input and temperature.

Overall, the climate system is responding to spatially and temporally varying energy input. The oceans and atmosphere, each being fluids of different mechanical and thermal inertial

---

84 At sea level, the density of water is 1,000 times greater than air and the heat necessary to raise the temperature of water 1°C is approximately four times that needed to raise a similar mass of air the same amount. The oceans are considered to be the 'flywheel' of the climate system.

characteristics, have a non-linear response to the seasonal forcing. They also interact with each other through exchanges of heat, moisture and momentum across the air-sea interface. As a consequence, the out of phase forcing between the hemispheres by seasonally varying solar radiation imposes complex variability on the climate system. The circulations of the atmosphere and the oceans are a response to the seasonal variability of solar radiation and the climate system is always working towards top of the atmosphere energy balance. However balance, when achieved, is only transient.

The Oceans

The oceans, after the atmosphere, form the second fluid of the climate system. Their circulations are responsible for between 10 and 15 percent of the total energy transported from the tropics to the polar regions each year. Despite this relatively small proportion of the total energy transport, the influence of the oceans cannot be ignored because they act as the flywheels of the climate system. The enormous mass of the major oceans mean that, once initiated, surface currents and other planetary scale ocean circulations tend to persist and continue their transport of energy around their respective basins.

The oceans also have an enormous thermal capacity when compared to the atmosphere. The scale of the different thermal capacities of the oceans and atmosphere can be gauged from the fact that, for each unit of area, the upper 3 metres of the ocean has the same thermal capacity as the atmosphere above. The overall mass of the oceans and their thermal capacity underscores that the oceans are a major energy reservoir and the flywheels of the climate system.

The enormous thermal capacity of the oceans ensures that apparently subtle changes to their characteristics can have a dramatic impact on the atmospheric circulation and regional climates. For example, El Niño events, involving anomalous warming of the eastern equatorial Pacific Ocean, displace global weather patterns from their usual locations and cause regional droughts, flood rains and severe storms[85]. The changes in the circulation of the tropical Pacific Ocean that are associated with El Niño events, and the consequent variations to global sea surface temperature patterns, are a primary contributor to the internal variability of the climate system on interannual timescales.

---

85 WMO, 1999. The 1997–1998 El Nino event: A scientific and technical retrospective. No 995.

Despite the differences in mass and density, the atmosphere nevertheless does have a strong ongoing influence on the oceans through wind stress and generation of wind-driven currents in the surface layer. The prevailing easterly winds maintain westward directed currents over the tropical oceans while the prevailing westerly winds of middle latitudes also contribute to the subtropical gyres of each hemisphere. The speeds of the ocean currents are low when compared to the motions of the atmosphere and it is the overall mass flow in the circulations and the temperature differences between the warm poleward directed and cooler equatorward directed boundary currents that ensure overall energy transport.

The Southern Ocean is a continuous water body that surrounds Antarctica and is generally in the latitudes of the southern hemisphere westerly winds. The persisting action of the westerly winds on the ocean surface generates the Antarctic Circumpolar Current, a current that has important ramifications for the climate of Antarctica and the southern hemisphere generally. The Antarctic Circumpolar Current is a barrier to warm water of subtropical origins washing the shores of Antarctica. The dynamics of the wind action also promote upwelling of cold water from the deeper ocean that accentuates cooling of ocean surface temperatures in the vicinity of Antarctica.

There are also other properties of the oceans, in addition to those related to energy and its transport, that are important when considering the role of the oceans in the climate system and its variability. Salinity, in combination with temperature, is a factor that regulates local density. It is suggested that variations in density are a primary cause of ocean convection and local overturning (or ventilation) of water masses. The density differences cause the more dense water mass to sink under the less dense one when water masses converge.

Dissolved gases, including carbon dioxide, are important for the ocean biosphere. The gases are also transported within the ocean circulation and are exchanged with the atmosphere. Gases, including carbon dioxide, are more soluble in colder water than warmer water and there is a tendency for gases to be released to the atmosphere over regions of tropical upwelling but to be extracted from the atmosphere where cold polar water is sinking to depth. The relative rates of release and uptake over the various regions of the globe are one factor governing their respective atmospheric concentrations. Dissolved carbon dioxide is also extracted from the ocean during biological production and locked in skeletal debris accumulating on the ocean floor.

The hydrological cycle is important to the oceans because of the influence of evaporation, precipitation and river runoff on the salinity, and thus density, of the surface layer. Precipitation, being fresh water, tends to dilute local salinity and make the ocean surface layer more buoyant. In contrast, evaporation of seawater tends to increase local salinity and density. Globally, maximum surface layer salinity occurs across subtropical latitudes where evaporation is high and exceeds precipitation. The waters over central parts of the ocean gyres of the North and South Atlantic oceans, the North and South Pacific Oceans and in the seas bordering the dry Arabian Peninsula are thus regions of maximum salinity. In contrast, the local surface salinity across equatorial regions and middle latitudes is constantly being diluted because there is excess precipitation over evaporation.

The outflow of fresh water by many major river systems has a readily discernible regional impact on surface salinity. The Gulf of Guinea, the Bay of Bengal, the Atlantic Ocean off the northeast coast of Brazil and the northern Gulf of Mexico all have reduced surface salinity because of the discharge of freshwater from major river systems. The freshwater river runoff floats above the more saline seawater as it mixes. A number of regions, including the equatorial eastern Pacific Ocean and the coastal regions of northeast Asia, northeast North America and southeast South America, have reduced salinity because of wind-driven upwelling of deeper water with lower salinity.

The various exchanges of latent energy that take place within the hydrological cycle are important to the overall energy cycle of the climate system but are especially important in the context of long-term climate variability during the expansion or contraction of polar ice sheets. Latent energy taken from the oceans during evaporation of water vapour is a primary source of energy for driving the atmospheric and ocean circulations. In the context of climate change this energy transfer is not normally considered important because of the relatively short residence time of water vapour in the atmosphere. Based on the average amount of water vapour in the atmosphere and the average evaporation and precipitation rates, water vapour cycles through the atmosphere in less than two weeks. That is, within two weeks the latent energy extracted from the oceans during evaporation is released to the atmosphere during precipitation. If the precipitation is as rainfall then the water finds its way back to the oceans and there is a match between the energy lost by the ocean and the energy gained by the atmosphere. Moreover, latent energy gained by the atmosphere partly compensates for net radiation loss to space by the troposphere.

The latent energy budget does, however, become crucial when precipitation is as snowfall. The winter expansion of the polar snowline represents a temporary extraction of additional latent energy by the climate system. This is because when water vapour sublimates to snow there is a release of latent energy to the atmosphere that is in excess of the latent energy extracted from the oceans during evaporation. The polar troposphere is cooling during wintertime and the latent energy offsets radiation to space. In summer, latent energy is taken up, either from the atmosphere or absorption of solar radiation. It should be noted that if the temperature of the ice or snow surface is below 0°C then there is an additional barrier to snow ablation. At these low temperatures the latent energy needed to directly evaporate ice, rather than melt it, is approximately 8 times greater than the latent energy needed for melting at 0°C.

Expansion of the polar ice sheets during long-term snow accumulation represents a systematic transfer of energy from the oceans to the atmosphere and thence loss to space by radiation. In the process there is also a transfer of water mass from the oceans to the polar ice sheets, and sea levels fall. The thickening of ice sheets is favoured when temperatures are maintained below 0°C because of the significant increase in energy needed for direct evaporation over the energy required for melting. We will return to the issue of why the 0°C temperature is important for the stability of polar ice sheets in our later discussion on internal variability of the climate system. Suffice now to note the ongoing loss of latent energy from the oceans to space, and the annual cycle of water mass exchange between the oceans and the polar ice sheets that is integral to the poleward transport of energy.

There is also planetary scale overturning of the ocean mass in addition to the wind-driven currents that are mainly observed in the surface layer. A tendency for overturning is promoted by the density difference in the surface layer between the warmer tropical waters and the colder polar waters. However, over most of the oceans, especially over tropical regions, the vertical column of water is stratified, with warm water (heated by solar radiation) overlying the colder deep waters. It is only in the region of near-freezing polar waters that salinity and temperature combine, aided by surface winds, to break down the vertical stratification.

The above brief outline is of some of the known characteristics of the oceans that are important for climate and its variability. However, when compared to the atmosphere, observations that are necessary to specify the oceans' structures, their circulations, and variations over time are sparse. Surface observations, especially of sea surface temperature and the tides, comprise the longest ocean records. As with meteorological data, development of instrument technology has progressed and improved the accuracy and frequency of observations. There is also an increasing use of automated instruments. However these more recent developments cannot overcome the lack of data from the past.

Sub-surface ocean data, in particular, are sparse because they have been expensive to acquire. Specially equipped research vessels and dedicated voyages have been needed to gather sub-surface ocean data at specific localities of interest or across ocean basins. The research programmes, often involving international cooperation, have yielded significant knowledge about the general thermal and salinity properties and local currents across the ocean basins. Such internationally coordinated research programmes continue to be essential for gathering a range of oceanographic data and for better understanding ocean processes and the variability of ocean circulations.

The irregular measurements at most localities mean that the nature of changes in properties over time is often difficult to characterise. Where there are only a limited number of observations at a locality the data give only limited insights into how such properties vary seasonally, from year to year, or over longer timescales. For example, a series of measurements of sub-surface properties were made across the Indian Ocean at about latitude 32°S during 1936, 1965, 1987, 1995 and 2002 during research voyages. These data show that upper thermocline waters (in each case, the layer between the 10°C and 17°C) apparently changed little from 1936 to 1965, freshened from 1965 to 1987, and since 1987 have become more saline. These fluctuations demonstrate substantial variability in sub-surface water properties but the infrequent observations tell us little about the real frequency and range of variability. It is also of note that the later increase in salinity is at odds with computer models that identified the earlier freshening as due to global warming[86].

Over recent decades there has been an expansion of systematic ocean observations. Expendable instruments, deployed from merchant vessels, provide frequent observations of temperature and salinity at depth along the main shipping routes. Satellites have provided spatial representations of sea surface temperature and, more recently, sea level. As part of the internationally coordinated Tropical Oceans Global Atmosphere (TOGA) research programme a network of more than 70 tethered buoys is anchored to the seafloor across the equatorial Pacific Ocean which continuously measure changes in temperature, salinity and ocean currents at various depths. A similar array of tethered buoys is being deployed across the equatorial Atlantic Ocean and is expected to provide knowledge about that ocean and its variability. Other tethered buoys provide sub-surface information from regional seas.

86 Bryden, H.L., E.L. McDonagh and B.A. King, 2003. Changes in water mass properties: oscillations or trends. Science, vol 300, pp2086–2088.

More recently, several thousand submersible drifting buoys have been released across the oceans. These ARGO buoys are expected to make observations at varying depths over an extended period of time. Periodically the buoys will float to the surface in order to relay their data via satellite communications. These buoys, reporting regularly from the world oceans, are expected to greatly enhance knowledge about sub-surface characteristics and their variability.

Knowledge about the extent of ocean variability and the interactions with the overlying atmosphere will expand and improve in time as the systematic observations accumulate and analyses of the structures and circulations are carried out. In the meantime, it is a dangerous presumption to ignore the potential long-term impacts of the oceans on climate. It is only relatively recently that it has been established that El Niño events occur on interannual timescales. However recent events underscore that these transient climate events cause regional level social disruption, economic loss and environmental degradation around the world. It is only through current research that decadal length drought, such as the North American 'Dust Bowl' of the 1930s, has been established as likely linked to persisting ocean surface temperature anomalies. The timescale of overturning of the oceans is of the order of 1,000 years and, consequently, significant climate variability on centennial to millennial timescales, as has been clearly documented in the Medieval Warm Period and Little Ice Age of the Northern Europe-North Atlantic region, cannot be discounted.

Surface temperature and upwelling

Solar radiation penetrates into and is absorbed within the surface mixed layer of the oceans. The energy of the solar radiation accumulates within the layer until it is transferred through the sea-air interface to the atmosphere. We can consider the surface mixed layer of the oceans as an energy reservoir for the climate system, constantly being replenished by solar radiation but losing energy to the atmosphere.

Local sea surface temperature is a primary control over the rate of energy transfer from the oceans to the atmosphere. First, temperature controls the rate of longwave radiation emission from the surface. Secondly, the rate of heat conduction from the ocean surface to the atmospheric boundary layer is regulated by the temperature gradient between the sea surface and the atmosphere. Thirdly, the evaporation (and latent energy transfer to the atmosphere) increases with the saturation specific humidity at the ocean surface, and the latter increases with temperature. The local magnitudes and global pattern of sea surface temperature are crucial for regulating the exchange of heat and moisture between the oceans and the atmosphere.

Local solar radiation is the primary control over local sea surface temperature. As a consequence, the global pattern of ocean surface temperature broadly follows the pattern of annual solar insolation. The map of annual average sea surface temperature (see Figure 15) shows that the maximum surface temperatures are generally in the vicinity of the equator and temperatures decrease toward the polar regions. Over middle latitudes the surface temperatures have an annual cycle that reflects the seasonal variations of solar insolation and wind patterns[87].

Surface currents, especially those associated with the major gyres of the subtropical oceans, also transport warm and cold water and modify the sea surface temperature pattern. Over polar regions the permanent ice maintains open water temperatures near 0°C all year round, despite a large range for the annual cycle of solar insolation, including permanent darkness for a period of winter and long daylight hours during summer. The seasonally changing solar insolation over polar regions is reflected as seasonal variations of sea ice extent.

Surface winds modify many of the seasonal and long-term characteristics of the surface mixed layer and sea surface temperature. In some regions the actions of the surface wind determine unique local characteristics. Wind speed promotes turbulent mixing in the atmospheric boundary layer and stronger winds increase the rate of heat conduction and evaporation from the ocean. Winds also promote wave action on the ocean surface that mixes seawater properties through the surface layer. In some regions the wind flowing over the ocean surface promotes upwelling of cold water from beneath the surface layer. Overall, winds promote downward mixing of energy in the surface layer and a transfer of energy from the oceans to the atmosphere. Wind stress also transfers momentum from the winds in the atmospheric boundary layer to the underlying ocean and sets up wind-driven surface currents. The interaction between the surface wind field and the oceans combines to shape the characteristics of the surface layer, including its surface temperature and salinity.

The thermocline is a narrow layer with a strong vertical temperature gradient that separates warmer surface waters from the cold waters below. The thermocline is generally within about a hundred metres of the surface. Although the actions of winds and breaking waves cause turbulence and mixing, the surface layer generally remains stratified, with warmest water at the surface, because of the ongoing absorption of solar radiation.

87   Harrison, D.E. and N.K. Larkin, 1998. El Niño/Southern Oscillation sea surface temperature and wind anomalies. Rev. of Geoph. 36(3) 353–399.

Over middle latitudes the vertical temperature gradient within the surface layer varies with the seasons because of the variation in solar heating between winter and summer. The layer is heated by the relatively strong solar insolation during summer and a vertical temperature gradient is established, with the warmest temperature near the surface. The temperature gradient represents a stratification that inhibits mixing through the layer. Solar insolation is at a minimum during winter and the layer loses heat across the sea-air interface, especially by conduction to the atmosphere. Over the western margins of the North Pacific and North Atlantic Oceans the heat loss is particularly strong because of the very cold winds blowing off the adjacent continents. There is strong radiation loss at the ocean surface during winter and turbulent mixing transports heat upward to the surface. The mixed layer slowly cools from above and a constant temperature layer develops at the surface.

Winds exert frictional stresses on the ocean surface and there is a transfer of momentum from the atmosphere to the ocean surface. Although the inertia of the atmosphere is much less than that of the oceans, the effect of the persisting winds and ongoing transfer of momentum is to set up wind-driven (or wind-drift) currents. The effect of the earth's rotation is to cause the direction of the wind-driven ocean current at the surface to diverge from the direction of the prevailing wind. The deviation of the direction of the ocean current from the wind direction is called the Ekman Drift. In the northern hemisphere the Ekman drift is to the right of the wind direction and in the southern hemisphere the divergence of the wind-driven current is to the left.

There are a number of regions of the world's oceans where the prevailing wind blows parallel to the coast, or is directed offshore. In these regions, the wind-driven current is also directed offshore and replacement water is drawn up from below the surface. If the water drawn up is from sufficient depth then the thermocline is raised to shallow depths. The action of the wind on the ocean that draws water upward from deeper layers is called upwelling. Where the upwelling is sufficiently strong, cold water from below the thermocline is entrained into and mixes with the surface layer to lower the sea surface temperature. As can be seen from Figure 15, there are a number of regions over the oceans where there are relatively strong departures from the general pattern that has sea surface temperatures decreasing towards the poles. Many of the cooler regions can be attributed to local upwelling of cold water.

The main locations of tropical and subtropical upwelling are on the eastern margins of the subtropical oceans. The anticyclonic surface winds of the subtropical atmospheric high pressure systems provide energy for the ocean gyres but the winds blowing parallel to the coast or offshore on the eastern side also maintain upwelling. Principal areas of ocean upwelling are the California Current of North America, the Peru Current of South America, the Canary Current of Northwest Africa, the Bengueta Current of Southwest Africa and the Somali Current of the Horn of Africa. The upwelling that takes place offshore from the Horn of Africa is seasonal. The southwest winds of the Asian summer monsoon circulation cause an offshore Ekman drift away from the African coast and upwelling at the coast.

Upwelling also occurs where middle latitude westerly winds during winter blow offshore from Asia and North America. Over the western margins of the North Pacific and North Atlantic Oceans sub-surface waters that are both cold and less saline are brought to the surface. The upwelling over the North Atlantic Ocean is particularly important because of the potential for the less saline water to mix with and freshen northward flowing surface water of the Gulf Stream.

The eastern Pacific Ocean is unique because of the upwelling and cooling of sea surface temperatures that extends from the South American coast along the equator to the central Pacific Ocean. Winds with a prevailing easterly component blow across the equator. However, the easterly winds south of the equator cause an Ekman drift in the surface water to the left away from the equator and the winds north of the equator cause an Ekman drift to the right, also away from the equator. At the equator there is a critical situation with the Ekman drifts on each side having opposing directions. Therefore, because the surface drift is away from the equator on each side, there is strong upwelling of colder sub-surface water. The upwelling is initiated offshore from the South American coastline and extends out into the equatorial central Pacific Ocean. Variations in the speed of the easterly winds and consequent upwelling of cold water is one of the reasons for the El Niño-La Niña sea surface temperature variability of the eastern equatorial Pacific Ocean.

The persisting westerly winds blowing over the Southern Ocean are the energy source for the Antarctic Circumpolar Current. The Antarctic Circumpolar Current also has an Ekman drift in the surface layer and surface water flows across the Current and away from the Antarctic continent. The Ekman drift also causes upwelling of cold Antarctic Intermediate Water on the southern flank of the Circumpolar Current. The upwelling water is very saline and near freezing. However local snowfall and the melting of sea-ice and icebergs from the Antarctic continent act to freshen the surface water near the Antarctic continent.

The circumpolar current and the upwelling cold water at about latitude 50°S have an important role in isolating and regulating the climate over Antarctica. The continuous flow of the Antarctic Circumpolar Current around the continental margin prevents incursions of warmer subtropical waters that might otherwise modify the climate. Under the influence of the prevailing westerlies and the Ekman Drift, the Antarctic surface waters move northward and freshen by precipitation. At the Antarctic convergence, the near freezing Antarctic waters meet the more temperate waters of the Southern Ocean that may be 15°C warmer. The Antarctic waters sink beneath the warmer waters to a depth of about 900 metres and move northward as Antarctic Intermediate Water[88]. Protected from incursions of warmer water, and in a region of ongoing net radiation deficit, the Antarctic continent is dependent on atmospheric transport of energy to maintain its average surface temperature.

Ocean overturning – the thermohaline circulation

The ocean surface layer is generally warmer and less dense than the deeper oceans. Turbulent mixing by wave actions does distribute heat downward but the depth of this influence is restricted to a about a hundred metres. The vertical mixing that occurs within the surface layer sharpens the boundary between the surface and deeper cold waters and causes a strong vertical temperature gradient over the narrow layer of the thermocline. The very strong stratification of the thermocline marks the boundary between the surface layer and the deeper water. The thermocline, with stable stratification, is a barrier to the vertical exchange of properties, including energy, dissolved salts and momentum.

Despite the overall stratification, the oceans have a planetary scale overturning, often called the thermohaline circulation[89]. The rate of overturning is known to have varied on timescales of centuries to millennia but the reasons for the overturning and its variability are controversial[90]. The thermohaline circulation provides continual replenishment of cold bottom water to the deep oceans and the global pattern of the circulation is well known. The two principal regions of cold bottom water formation are the sub-Arctic waters of the North Atlantic Ocean and the near-shore waters of the Southern Ocean surrounding Antarctica.

Although the reasons for the thermohaline circulation are controversial, what is observed is that, over the polar regions during

---

88   Fifield, R., 1987. International research in the Antarctic. Oxford University Press.
89   Stommel, H., 1961. Thermohaline convection with two stable regimes of flow. Tellus, vol 13, pp131–149.
90   Wunsch, C., 2002. What is the thermohaline circulation? Science, vol 298, pp1179–1180.

winter, cold surface water of high salinity sinks to the deep ocean and produces bottom water. The sub-surface thermocline disappears because of the uniformly cold temperatures of the water; however, vertical density gradients in the water column tend to form because of the freshening of surface water[91] that is attributable to precipitation, ice melt and coastal upwelling.

The North Atlantic Gulf Stream transports water of high salinity northward to the sub-Arctic region. In spite of a tendency for the surface waters to freshen, because of local precipitation, ice melt and mixing with upwelling water forming off the Labrador coast, the surface waters cool to near freezing and the density increases. The magnitude of the density of the sub-Arctic surface waters is dependent on the salinity of the source waters in the tropical Atlantic Ocean and the wintertime cooling. The speed at which the Gulf Stream transports saline water into the sub-Arctic region is a factor assisting production of the high density that is achieved by the surface waters. The strong Gulf Stream flow, with its saline water, tends to minimise the freshening effect on surface waters by precipitation, ice melt and upwelling.

The local salinity of surface water is enhanced during the formation of sea ice. This is because salt is expelled into the surrounding water as seawater freezes and sea ice forms. The salinity and density of water increases under thickening sea ice. During their respective winters, the pack ice of the Arctic Ocean thickens and extensive sea ice forms around Antarctica so that there is a seasonal increase in the salinity and density of the water in the surface layer. The cold and dense surface water that forms in each polar region sinks to the ocean floor and the bottom water slowly spreads equatorward. The rate of cooling and the salinity of the surface waters of each polar region affect the respective rates of formation of North Atlantic Bottom Water and Antarctic Bottom Water.

High salinity and cooling are necessary conditions to ensure that the surface waters of polar regions become dense enough to sink under the influence of gravity. However there is a view that, in order for the cold deep waters to return upward against the stable stratification that exists almost everywhere, there is a need for work to be expended. The controversy is about whether the source of energy for the overturning circulation is wind action or whether density differences (ie, negative buoyancy due to cold saline polar waters) is sufficient. The view that the planetary thermohaline circulation is sustained primarily by surface winds is supported by theoretical studies that suggest turbulent

91   Sigman, D.M., S.L. Jaccard and G.H. Huang, 2004. Polar ocean stratification in a cold climate. Nature, vol 428, pp59–63.

mixing is required to carry the colder waters upward across the density gradients of the thermocline.

The strength and stability of the thermohaline circulation are important when considering climate variability and sudden climate shifts. The formation of North Atlantic Bottom Water is a crucial component of the ocean conveyor proposed by Professor Wallace Broecker of Lamont-Doherty Earth observatory, USA during the 1980s. The ocean conveyor represents the movement of water between the oceans, including its overturning. From its formation in the North Atlantic Ocean, the cold bottom water spreads around the globe in the deep oceans and finally emerges at the surface in the areas of upwelling, particularly of the subtropical oceans. At the surface there is a net movement of water from the Pacific Ocean, through the Indonesian Archipelago and Indian Ocean, around southern Africa, northward through the Atlantic Basin and finally returning to the sub-Arctic region of the North Atlantic where formation occurs. The mean time of overturning of the thermohaline circulation through the ocean conveyor is about 1,000 years.

The contemporary distribution of regions of bottom water formation is not necessarily unique. As noted, high salinity and near freezing temperatures are necessary and it is the prevailing atmospheric and ocean circulations that favour the North Atlantic and Southern Ocean. The North Pacific Ocean is currently not a favoured region and one reason cited is the relatively slow circulation of the North Pacific Ocean gyre that allows freshening in the most northerly latitudes through prolonged exposure to precipitation and upwelling off the North Asia coast. The faster Gulf Stream causes saline surface waters of the tropical North Atlantic Ocean to reach high latitudes more quickly than those waters of the tropical Pacific Ocean. There may also be a degree of feedback so that the more rapid circulation and increased salinity of the North Atlantic are enhanced by bottom water formation.

Experiments with simple computer models indicate that, depending on the initial conditions, stable circulations can be developed with sinking in the North Atlantic, the North Pacific, both oceans, and neither ocean[92]. A more complex computer model has been used to simulate the climate with a changing input of freshwater into the North Atlantic[93]. The millennium scale variation in freshwater input produced an abrupt reduction in the magnitude of bottom water

92 Marotzke, J. and J. Willebrand, 1991. Multiple equilibria of the global thermohaline circulation. J. Phys. Ocean. v 21 pp1372–1385. For a discussion see: Held, I.M., 1993. Large-scale dynamics and global warming. BAMS, vol 74, No 2 pp228–241.

93 Ganopolski, A. and S. Rahmstorf, 2001. Rapid changes of glacial climate simulated in a coupled climate model. Nature, vol 409, pp153–158.

formation and a corresponding abrupt change in climate characteristics, including to colder temperature over Greenland and colder sea surface temperature over middle latitudes of the North Atlantic Ocean.

The computer model simulations identify abrupt changes to the climate in response to changing salinity over the North Atlantic Ocean. However, the limitations of even the most complex computer models, and our inability to resolve how the prevailing climate compares to climatic conditions near such abrupt changes, precludes realistic predictions of future climate shock[94].

Freshening of the North Atlantic surface waters, for example by increased precipitation and coastal runoff, will tend to slow the thermohaline circulation and the northward transport of heat. The rate of cooling is also affected by changes in the net longwave radiation loss at the surface and by the strength of the prevailing surface winds. Surface wind speed is one of the factors that regulate eddy transfer of heat and latent energy to the atmosphere. A warming of the overlying air or an increase in greenhouse gas concentrations can act to reduce net longwave radiation at the surface and hence reduce the rate of cooling of the surface waters. Persisting changes to any one, or a combination of the meteorological processes in the atmospheric boundary layer can vary the rate of formation of North Atlantic Bottom Water. There is some evidence that the thermohaline circulation has slowed over the past 50 years.

The strength of Circumpolar Current of the Southern Ocean is a factor that affects the rate of formation of Antarctic Bottom Water. Westerly winds south of 40 degrees latitude not only impart momentum to the current but, because of the Ekman Drift away from the continent, they also contribute to upwelling of cold Antarctic Intermediate Water on the southern flank of the Circumpolar Current. The upwelling water is very saline, near freezing and a ready source for bottom water formation. In opposition to the upwelling, falling snow and the melting of sea-ice and icebergs act to freshen the surface water near the Antarctic continent. Any change in the atmospheric circulation that varies the westerly wind strength or the distribution of rainfall will also have an impact on the rate of Antarctic Bottom Water formation.

Previous abrupt climate change events have been identified from a range of palaeoclimatic data, including the ice cores retrieved from Greenland and ocean sediment cores retrieved from various Atlantic Ocean locations. The records indicate that there have been times of near synchronous change in North Atlantic Bottom Water formation,

94 Bard, E., 2002. Climate shock: abrupt changes over millennial time scales. Physics Today, December pp32–38.

temperature changes of about 10°C over Greenland and several degrees change in sea surface temperature over middle latitude. These records represent climate characteristics of the locality from which they were obtained and are neither sufficiently detailed nor spatially representative to describe global patterns preceding the abrupt change. The limited amount of analysed data mean that it is not possible to identify whether bottom water formation and climate vary synchronously on shorter timescales or whether the linkage is only at a threshold of major abrupt change.

The global reach of climate variations associated with fluctuations in North Atlantic Deep Water formation is also not well established. Some palaeoclimatic evidence, such as from a Taylor Dome ice core[95], in the west Ross Sea sector of Antarctica, suggests a synchronicity with Central Greenland climate fluctuations over the last glacial epoch but this is not reproduced in data from other Antarctic locations. Part of the problem with establishing synchronicity of climate events

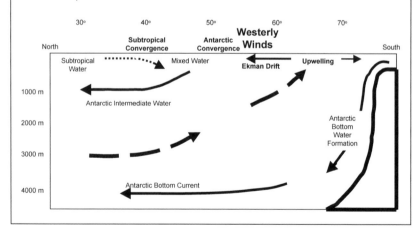

Meridional Overturning of the Southern Ocean
- Westerly winds drive the Antarctic Circumpolar Current and there is an Ekman Drift towards the equator in the surface layer.
- Upwelling occurs on the poleward side of the Antarctic Circumpolar Current
- Very cold equatorward flowing water of the Ekman Drift plunges under the subtropical waters at the Antarctic convergence.
- Antarctic Bottom Water forms in winter as salt is expelled during freezing and increases the water density under the coastal sea ice.

Figure 21: Schematic representation of meridional overturning in the Southern Ocean

95 Steig, E.G., E.J. Brook, J.W.C. White, C.M. Sucher, M.L. Bender, S.J. Lehman, D.L. Morse, E.D. Waddington and G.D. Clow, 1998. Synchronous climate changes in Antarctica and the North Atlantic. Science, vol 282 pp92–95

identified over the two polar regions arises from difficulties in precisely dating the events in the ice cores. The problem of accurate dating comes about because air trapped in the porous surface snow and ice diffuses downward until bubbles reach a depth in which they are trapped. The difference between the age of the ice layer and the age of the trapped air bubbles may be several hundred years[96] and varies according to the accumulation rate of surface snow.

There is evidence from during the Holocene of significant climate fluctuations affecting the North Atlantic but data from elsewhere do not always corroborate. The lack of corroboration for global scale synchronicity may reflect ambiguity in the response of the selected proxies as much as no climate fluctuation having occurred. Some widely separated ocean-floor sediment cores are claimed to have coherence with Greenland climate fluctuations during the Holocene, including with the Medieval Warm Period and the Little Ice Age. The cores include those from Bransfield Basin adjacent to the Antarctic Peninsula[97], from the Bermuda Rise in the western Atlantic Ocean[98], and from off the coast of West Africa[99]. There is much evidence of significant climate fluctuations over land areas but generally these are neither synchronous with the North Atlantic fluctuations nor are the temperature and humidity fluctuations coherent[100].

There is also evidence that bottom water formation around Antarctica[101] has undergone substantial change over recent centuries. Properties of well-mixed deep water in the circumpolar current around Antarctica suggest that, over the past 800 years, formation has come almost equally from the North Atlantic and around Antarctica. However, analysis of concentrations of chlorofluorocarbons (industrial compounds not occurring naturally) dissolved in recently formed deep water suggest that the current formation rate of Antarctic Bottom Water is significantly less than the estimates for earlier times. The estimated current rate of formation of Antarctic Bottom Water is also several times less than the rate of formation of North Atlantic Bottom Water formation. Such a reduction in Antarctic Bottom Water

---

96 Stocker, T.F., 2002. North-south connections. Science, vol 297, pp1814–1815.

97 Kim, B-K., H.I. Yoon and C.Y. Kang, 2002. Unstable climate oscillations during the late Holocene in the central Bransfield Basin, Antarctic Peninsula. Quart. Res. vol 58, pp234–245.

98 Keigwin, L.D., 1996. The Little Ice Age and Medieval Warm Period in the Sargossa Sea. Science, vol 274, pp1504–1507

99 deMenocal, P. et al, 2000. Ibid.

100 See the earlier discussion on climate during the Holocene.

101 Broeker, W.S., S. Sutherland and T-H Peng, 1999. A possible 20th-century slowdown of Southern Ocean deep water formation. Science, vol 286, pp1132–1135.

formation is consistent with reported freshening of waters of the Ross Sea during observations over the past four decades[102].

Episodic changes in Antarctic Bottom Water formation that are not linked to external forcing processes point to internal ocean variability on the millennial timescale. The trigger for such variability is not obvious but it may relate to changing salinity of upwelling water or the rate of upwelling, which is related to westerly wind speeds. It is also not clear whether the recent slowdown has been in response to, or part of, the trend of global warming that has been occurring in concert with glacial retreat since the middle 19th century.

The El Niño Southern Oscillation (ENSO)

The El Niño Southern Oscillation, or ENSO, is a coherent set of ocean-atmosphere fluctuations on the interannual timescale. After the seasonal cycle, ENSO is the most dominant signal of climate variability. It is not a cyclic oscillation that fluctuates with a specific periodicity. However, statistics indicate that the time between a significant El Niño event is between four and seven years.

The primary indicator of ENSO is the interannual variability of sea surface temperature over the central and eastern equatorial Pacific Ocean. During an El Niño event, warm surface waters of the western Pacific Ocean are observed further eastward, the normal easterly Trade Winds of each hemisphere weaken and the deep tropical convection normally associated with the western Pacific warm pool shifts eastward. At the same time there is a weakening of the broad zonal overturning of the atmosphere (called the Walker Circulation) that has ascending air in the deep convective clouds over the tropical western Pacific Ocean and Asia, and compensating subsiding motion over the central and eastern Pacific Oceans.

The shift in the location of deep convection from the western to the central and eastern Pacific Ocean alters the intensity of the overturning Hadley Cells and influences the location and speed of the subtropical jetstreams. Further, the Rossby Waves in the westerly winds of the middle latitudes and the associated upper troposphere jetstreams are also affected. There are shifts in the preferred paths of seasonal weather patterns. Some regions encounter higher than normal frequencies of seasonal rain events and these regions often experience flooding of river valleys and other low-lying areas. In other regions the seasonal rains are reduced to such an extent that crops, pastures and water supplies fail.

El Niño events occur because of inherent instability in the interactions between surface winds and the ocean surface layer across

---

102 Jacobs, S.S., C.F. Giulivi and P.A. Mele, 2002. Freshening of the Ross Sea during the late 20th century, Science, vol 297, pp386–389.

the equatorial Pacific Ocean. The persisting Trade Winds establish a westward directed wind-drift current. We have already noted that the surface waters of the eastern equatorial Pacific Ocean are abnormally cool because of the upwelling that occurs off the South American coast and extends westward along the equator. As the surface waters are carried westward by the wind-drift current they are warmed by solar radiation.

The warmest waters of the equatorial Pacific Ocean are found to the west of the international Date Line (180 degrees longitude) where the waters of the wind-drift current accumulate as a warm pool. In addition to the surface temperature gradient across the Pacific Ocean there is also a rise in sea level towards the west that occurs because of both the changing temperatures of the ocean surface layer and of the action of the wind stress on the surface. Sea level is approximately 60 cm higher over the warm pool than off the coast of South America. The action of the wind stress is to deepen the thermocline in the region of the warm pool. The overall warmer waters and deeper thermocline of the west can be seen in the equatorial cross-section across the Pacific Ocean in Figure 22. Deep tropical convection is mainly over the western part where sea surface temperatures are greater than 28°C (see also the map at Figure 16).

The elevated sea level and deeper thermocline of the warm pool are forced by the wind stress of the easterly Trade Winds. When the easterly winds weaken there is a tendency for the sea level set-up to ease and for warm waters in the surface layer to shift eastward. However, an eastward shift in the 28°C boundary that separates the warm pool will also draw deep convection further eastward. When this happens, the easterly winds over the warm pool are further weakened as boundary layer wind patterns adjust to ensure airflow into the deep convection. Bursts of equatorial westerly winds over the warm pool, either associated with the Australian summer monsoon, eastward propagating weather disturbances from the region of Indonesia, or persisting tropical cyclones, have been observed in conjunction with an eastward movement of warm water from the warm pool.

A strengthening of the easterly Trade Winds and the associated stress on the ocean surface will cause the thermocline beneath the warm pool to deepen. Such a strengthening of the Trade Winds will generate a 'wave' on the thermocline that propagates eastward as a sub-surface warm anomaly on the thermocline – in much the same principle as ripples spread across a pond. The sub-surface warm anomaly of the wave reaches the coastal waters of the eastern Pacific Ocean after a little more than two months. The impact of the wave depends on the time of year that it arrives in the eastern Pacific Ocean. Between August and November there is strong upwelling and sea surface temperatures are only about 21–22°C. This temperature is too cold to support deep convection. However during March–April the sea surface temperatures are normally about 26°C. The arrival of a warm anomaly of about 2°C is enough to raise the sea surface temperatures sufficiently to support deep convection and interact with the atmospheric circulation of the troposphere.

There are two quite different effects from the fluctuations of the speed of the Trade Winds. An easing of the Trade Winds (or westerly wind burst) will initiate an eastward movement of warm water, while a strengthening, or pumping, of the Trade Winds will establish a sub-surface wave on the thermocline that propagates eastward. Both

The west to east equatorial cross-section of the upper 500 metres of the Pacific Ocean identifies the warmest waters at the surface accumulating over the western side.

The thermocline layer of strong vertical temperature gradients separate the warm surface layer from the colder waters at depth. The thermocline is deepest in the west and rises to near the surface in the east where strong wind-generated upwelling takes place.

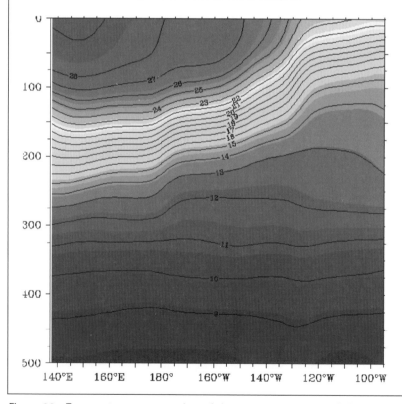

Figure 22: Temperature cross-section of the upper 500 metres of the equatorial Pacific Ocean

effects are an aberration from the normal cross-Pacific set-up but neither always results in an El Niño event. There is inertia in both the ocean surface current and the atmospheric circulation and work has to be expended to bring about change. It is necessary that the easing of the wind over the warm pool is sufficiently prolonged to enable the atmospheric circulation to respond to the new location of deep convection and associated pattern of tropical forcing. As we have noted, it is also essential for the sub-surface wave on the thermocline to be of sufficient magnitude and for its emergence in the east to be at a time that is conducive to deep convection.

The most intense El Niño events are those associated with eastward movement of warm surface water, such as in 1997–98[103]. During that event, as with the intense 1982–83 event, warm water spread eastward across the eastern equatorial Pacific Ocean and northward and southward along the Pacific coast of the Americas. Warm water above 28°C persisted over the central Pacific Ocean east of the Date Line for 12 months from May 1997 through May 1998. Although abnormally warm surface waters were observed over the eastern Pacific Ocean at the same time and persisted longer, the sea surface temperatures were only above 28°C and able to support deep convection from January to May of 1998.

ENSO profoundly affects the global climate on the interannual to multi-year timescale. Enhanced energy and momentum are exchanged between the equatorial ocean and the atmospheric boundary layer during an El Niño event and the tropical pattern of sea surface temperature changes. The changing spatial patterns of deep tropical convection during an ENSO cycle are the primary links to changes of the global atmospheric circulation, and hence the climate impacts. The longitudinal displacement of the focus of tropical convection is reflected in modification to the Rossby Waves and subtropical jet streams of the upper troposphere. These in turn impact on the tracks and intensities of middle latitude weather systems[104].

The situation of relatively cold surface waters over the equatorial eastern Pacific Ocean has been dubbed a La Niña event. Although the waters of the central and eastern equatorial Pacific are cool and the easterly Trade Winds strengthen during the La Niña phase of ENSO, the climatic effect is not a direct opposite to the El Niño phase. The spatial distribution of deep tropical convection is generally similar to

103 WMO, 1999. Ibid.
104 Trenberth, K.E., G.W. Branstator, D. Karoly, A. Kumar, N-C. Lau and C. Ropewleski, 1998. Progress during TOGA in understanding and modelling global teleconnections associated with tropical sea surface temperatures. J. Geophys. Res., vol 103, pp14,291–14,324.

what is usually observed, except local rainfall frequency and intensity are enhanced over the islands of Indonesia and during the summer monsoon periods of Asia and Australia. Similarly, the Rossby Waves of the upper troposphere are near their normal patterns and, as a consequence, middle latitude weather systems are not disrupted from their normal tracks, but are more frequent and more intense than normal over these locations.

The El Niño phenomenon has its origins in variability of sea surface temperatures across the equatorial Pacific Ocean and there is generally a synchronous response, through processes not well understood, in the tropical Atlantic and Indian Oceans. All three tropical oceans warm and increase the rate of eddy transfer of heat and moisture from the ocean. The increased energy in the tropical boundary layer enhances deep convective activity and additional heat is available to the troposphere during major El Niño events[105]. Zonally averaged, the temperatures of the mid-troposphere, as measured by satellite microwave sounding units, rose slightly during the 1982–83 and 1987–88 El Niño events[106]. An extended record of satellite measurements that include corrections for orbital drift confirms the earlier analysis and also shows the mid-tropospheric warming during the major 1997–98 El Niño event[107].

Research, especially the internationally coordinated Tropical Ocean Global Atmosphere (or TOGA) programme of the Pacific Ocean carried out between 1985 and 1995 and following programmes, has expanded knowledge of ENSO. There are now systems in place to monitor the important characteristics of the ENSO pattern as they wax and wane. Although ENSO is far from being fully understood the characteristic pattern of global impact has been well identified and monitoring programmes do provide a basis for early warning. This knowledge is the basis for better planning, for developing appropriate response strategies and for mitigating some of the worst impacts, such as famine and disease in developing countries.

As more data are gathered, there is emerging information about the variability of other aspects of the circulations of the oceans beyond the interannual fluctuations of the tropical Pacific Ocean. Such variability includes expansion of the western Pacific Ocean warm pool over extended periods such as has occurred since the middle 1970s, coherent fluctuations of the sea surface temperatures over the North

105 Trenberth, K.E. et al, 2002. Ibid.

106 Spencer, R.W. and J.R. Christy, 1990. Precise monitoring of global temperature trends from satellites. Science, vol 247 pp1558–1562.

107 Christy, J.R., R.W. Spencer, W.B. Norris, W.D. Braswell and D.E. Parker, 2003. Error estimates of version 5.0 of MSU-AMSU bulk atmospheric temperatures. J. of Atmos. and Ocean Tech., vol 20, pp613–629.

Pacific Ocean on decadal timescales, and circulating temperature anomalies in the Antarctic Circumpolar Current. Although different forms of ocean surface temperature variability have been identified, the triggering mechanisms and their impacts on the atmospheric circulation and local weather patterns are yet to be confidently resolved.

Despite the rapid development in knowledge of the oceans' structures and their circulations there remain many unknowns about the longer timescales of variability and the interactions between the changing circulations of the atmosphere and the oceans. Knowledge of the characteristics of the longer timescales of variability will continue to be limited because of lack of data with sufficient precision on these timescales. As with global temperature and precipitation, some inferences and insights can be drawn from proxy data. Computer models constructed to represent the oceans' circulations will be deficient until knowledge of the variations and characteristics of ocean circulations on the decadal, centennial and millennial timescales is developed. There will be uncertainty in model performance because there are no suitable benchmarks against which to assess amplitudes of variability on these timescales.

Carbon dioxide and Climate Variability

The Milankovitch theory of regularly changing earth orbital characteristics became widely accepted as the primary cause for ice ages after the 1930s. This was despite the lack of a satisfactory theoretical framework linking the periodic orbital characteristics to changes in the earth's temperature. The Milankovitch theory gained stronger support when deep cores from the ocean floor and polar ice sheets were extracted and analysed. Although other more random variations were superimposed, it was possible to identify signals of about 100,000 years, 40,000 years and 22,000 years associated with the respective orbital characteristics of eccentricity, obliquity and precession.

Despite the prominence gained by Milankovitch during the 20th century, Arrhenius' greenhouse theory, which had drifted out of favour, was far from forgotten. Feedback between atmospheric carbon dioxide and the biosphere was proposed as a possible amplification factor for the Milankovitch orbital forcing. In addition, the quantity of coal being burned at the beginning of the 20th century in support of industry had led to suggestions, even at that time, that it might be a cause for global warming, although the magnitude was not then quantified. Arrhenius, being sensitive to the cold winters experienced in his native Sweden, is reputed to have commented that warming may not be altogether a bad thing.

By the early 1950s a carbon dioxide-climate linkage was again under investigation. Evidence was mounting that the earth had been warming and that atmospheric carbon dioxide concentration was also increasing. As a component of the 1957–58 International Geosphere Year, an internationally coordinated program was established to gather observations of atmospheric carbon dioxide concentration from a worldwide network. Mauna Loa, Hawaii was the first station in the network and commenced systematic observations in 1958. These data, and those from later stations added to the network, soon confirmed that atmospheric concentrations of carbon dioxide and other greenhouse gases being measured were increasing relatively rapidly.

The source of the increase in atmospheric carbon dioxide has been attributed primarily to burning of fossil fuels, including for energy generation, transport, reduction of limestone in cement manufacture, and a range of other economic activities. The distortion of the natural exchange between the atmosphere and biosphere by land-use changes has also been identified as contributing to increased carbon dioxide and methane concentrations in the atmosphere. Not only has the range of human activities increased the rate of emission of carbon dioxide to the atmosphere but the magnitude has also increased significantly since early in the 20th century.

The first attempts to quantify the potential impact that an increase in atmospheric concentrations of greenhouse gases would have on climate were made in the late 1970s. The computer models used to simulate the climate system were adaptations of those being developed for weather forecasting. The atmospheric component was coupled to simple representation of the oceans and polar ice sheets. Simulations were made both with contemporary carbon dioxide concentrations (the control) and with the effective atmospheric concentration of carbon dioxide doubled (Double CO2). The differences between the equilibrium global average surface temperature of the Double CO2 and that of the Control provided an assessment of the potential impact of the carbon dioxide increase. The experiment, repeated at several laboratories, supported Arrhenius's earlier conclusion that global surface temperature might be sensitive to atmospheric carbon dioxide concentration. The array of independent assessments gave the first quantitative estimates of what the impact of human activities on climate might be, but the range of estimates also pointed to the uncertainty of the predictions.

The Villach Conference Statement of 1985 was, therefore, confirmation that the concentration of atmospheric carbon dioxide in the atmosphere (and that of other greenhouse gases measured across the global network) was increasing and endorsement, based on the modelling evidence, that global temperature is sensitive to changing greenhouse gas concentrations. Over nearly two decades since the Villach Conference there have been significant advances in computer modelling of the climate system. The advances include better specification of the ocean component and of ocean circulations, better representation of a range of physical processes, and better coupling between the atmospheric and ocean components of the models. Notwithstanding these advances, a continuing wide variation in the responses to similar greenhouse gas forcing is observed between the models developed in different laboratories. These vari underscore the continuing uncertainty surrounding the com modelling methodology and projections of future climate.

The IPCC conclusions about the reasons for recent increases in near-surface air temperatures and the projections for future warming come from a flawed view of the climate system and still rudimentary computer models that do not yet represent the climate system adequately. The IPCC hypothesis gives undue emphasis to global radiation processes at the expense of regional radiation imbalance and energy transport processes necessary to bring about near global radiation balance. The computer models used to predict future global warming give emphasis to cloud-radiation and water vapour feedback processes that have not been quantified through atmospheric observations. Also, the models are not capable of an adequate representation of the flow of energy through the climate system from the tropics to polar regions.

Growth of atmospheric carbon dioxide

The atmospheric concentration of carbon dioxide has been accurately measured for nearly 50 years and data from many locations confirm that the concentration has increased over this period. Estimates from gas trapped in polar ice provide a record of the variation in concentration extending over more than 400,000 years. The consistent picture is of concentrations that varied between about 180 ppm (parts per million) during glacial epochs and 280 ppm during the warmer interglacial periods. Since 1800, when the concentration was 280 ppm, there has been a continuing increase in concentration that reached 367 ppm in 1999.

The record of carbon dioxide concentration (and other trapped gases) from ice cores is not altogether free from controversy. Polar ice is built up in annual layers as a result of the seasonal snowfall and each layer contains air trapped at the time of snowfall. The snow compacts with time as new snow falls but there is diffusion of air between the layers and the overlying atmosphere. It is only when a layer is buried and compacted sufficiently that vertical diffusion from the layer ceases. The gas trapped in bubbles of ice, therefore, is not an exact representation of properties in the atmospheric boundary layer at the time of deposition but a smearing of properties from earlier and later years, and even many decades if accumulation rates are low. We can be confident that the ice core record reflects the variability of atmospheric carbon dioxide concentration over long timescales but we cannot discount smoothing of the magnitudes of shorter timescale events.

The recent increase in the atmospheric concentration of carbon dioxide is shown graphically in Figure 23. The increase is attributed to such human activities as fossil fuel usage, industrial processes and forest clearing. From audits of fossil fuel production it has been estimated that anthropogenic activities caused the release of about 6.3 PgC/yr (Peta grams of carbon per year, where 1 Pg = 10^{15} grams or 1,000 million tonnes) during the decade of the 1990s. Of the total anthropogenic emissions, about 3.3 PgC are estimated to have been retained in the atmosphere while the oceans and the terrestrial biosphere took up the remainder. The annual anthropogenic emissions are small compared to the total mass of carbon in the atmosphere (730 PgC). The annual anthropogenic emissions are also relatively small when compared to the exchange between the atmosphere and the terrestrial biosphere (120 PgC/yr) and the oceans (90 PgC/yr).

Within the terrestrial biosphere growing trees take carbon dioxide from the atmosphere during photosynthesis while decaying vegetation oxidises carbon and returns carbon dioxide to the atmosphere. The magnitude of this annual exchange and its impact on the regional

Carbon dioxide concentration in the atmosphere (ppm) has increased during the 1990s. The representation shows that concentrations are highest in the northern hemisphere. There is also a large annual cycle superimposed on the signal that has maximum amplitude over the high latitudes of the northern hemisphere. The annual cycle follows the pattern of summer growth (photosynthesis draws carbon dioxide out of the atmosphere) and winter dormancy and leaf matter decay (oxidation to carbon dioxide) over the middle and high latitude deciduous forests.

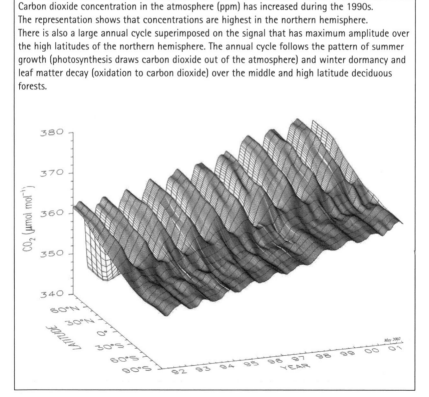

Figure 23: The increasing concentration of carbon dioxide in the atmosphere

concentration of carbon dioxide in the atmosphere are also evident in Figure 23. The extensive forested lands of the temperate and high latitude lands of the northern hemisphere are characterised by tree species that have a massive growth spurt during the warmer months when photosynthesis extracts carbon dioxide from the atmosphere. The trees are dormant during the colder months and decaying leaf material, etc., oxidises to carbon dioxide and the atmospheric concentration of carbon dioxide increases. The annual growth pattern of the trees of the boreal forests is reflected in the annual cycle of the atmospheric carbon dioxide record.

The growth rate of some plant species is currently limited by atmospheric carbon dioxide and these species grow more vigorously as the atmosphere is enriched. For other plants, it is their current adaptation to local sunlight, temperature, groundwater, nutrients, or a combination of these, that regulates photosynthesis and growth rates. It is these internal and environmental factors that prevent more vigorous growth, and hence more carbon dioxide uptake, in a carbon dioxide enriched atmosphere. Over recent years the terrestrial biosphere has taken up part of the anthropogenic carbon dioxide emissions but there is no consensus that this proportion might increase if anthropogenic emissions were to stabilise.

Over the oceans carbon dioxide is being vented to the atmosphere from regions of upwelling cold waters, especially in the tropics as these waters are warmed at the surface. At the same time, carbon dioxide is absorbed in the cold waters of the high latitudes. The lowest concentration of atmospheric carbon dioxide is over polar regions of the southern hemisphere. This suggests that the near freezing surface water that sinks to the deep ocean as part of the thermohaline circulation might be the primary uptake for atmospheric carbon dioxide. However observations and model studies suggest that the primary uptake mechanism is the equator-directed Ekman Drift of the Antarctic Circumpolar Current (see Figure 21) that plunges under the warmer subtropical waters at the Antarctic convergence[108].

Empirical studies suggest that the venting and uptake of carbon dioxide by the oceans are in near balance and there is limited capacity for the oceans to take up additional carbon dioxide. It has been estimated that there has been an ocean-wide net uptake of anthropogenic carbon dioxide of about 1.6 and 2.0 PgC per year during the decades of the 1980s and 1990s respectively[109]. Further

---

108 Caldeira, K. and P.B. Duffy, 2000. The role of the Southern Ocean in uptake and storage of anthropogenic carbon dioxide. Science, vol 287, pp620–622
109 McNeil, B.I., R.J. Matear, R.M. Key, J.L. Bullister and J.L. Sarmiento, 2003. Anthropogenic CO_2 uptake by the ocean based on the global chlorofluorocarbon data set. Science, vol 299, pp235–239

uptake by the oceans is limited because of sea surface alkalinity and the slow overturning of the ocean.

In the context of the palaeoclimate records, especially the polar ice core analyses, the increase in atmospheric carbon dioxide concentration over the past century is generally recognised as being very unusual. During glacial periods the atmospheric carbon dioxide concentration analysed from ice cores did not respond to the marked temperature changes that took place over decade to century timescales although there were fluctuations of up to 20 ppm associated with the more long-lived events. The apparent lack of response may have been due to the lack of short timescale resolution already mentioned. Despite the tendency for smearing of the record, there is excellent agreement among various high resolution Antarctic ice cores covering the past 1,000 years and these show that the atmospheric concentration fell by about 8 to 10 ppm over the period of the Little Ice Age from 1280 to 1860.

An independent estimate of atmospheric carbon dioxide concentration has been made based on an ocean-floor sediment core from the western equatorial Pacific Ocean[110]. The fraction of an isotope of boron in the accumulated shell remnants is related to alkalinity and carbon dioxide concentration of the surface water; however, the slow accumulation rates restrict timing to near millennium intervals. The analysis identifies that the concentration of carbon dioxide in the surface waters increased markedly over the interval 14,000 to 18,000 years ago and reached values up to 100 ppm more than contemporary atmospheric concentrations identified from Antarctic ice cores. The differences in carbon dioxide concentration have been attributed to changing rates of upwelling and venting of gases over the eastern equatorial Pacific Ocean and point to atmosphere-ocean interactions, especially varying wind stress, having a role in regulating atmospheric carbon dioxide production.

The natural processes that will regulate future concentrations of atmospheric carbon dioxide cannot be predicted with any certainty. There are still major unknowns about how ocean and terrestrial biosphere processes regulate the exchange of carbon with the atmosphere, particularly at the level of detail necessary to resolve the impact of anthropogenic emissions. The expectation is that the uptake by the oceans and terrestrial biosphere will respond to higher atmospheric concentrations even though uptake rates will not increase sufficiently to take up all additional emissions. Natural chemical buffering and the slow overturning circulation are expected to inhibit

---

110 Palmer, M.R. and P.N. Pearson, 2003. A 23,000-year record of surface water pH and PCO2 in the western equatorial Pacific Ocean. Science, vol 300, pp480–482

additional ocean uptake; additional uptake by the terrestrial biosphere is already limited by other environmental factors, including surface temperature and precipitation distributions.

The prediction of future anthropogenic contributions to atmospheric concentrations of carbon dioxide is also cause for serious debate. The available reserves and accessibility of the various fossil fuel types, future efficiencies of various industrial and thermal processes, population growth and the development paths of different economies are all factors that need to be taken into account. To a large extent the past is a guide to the future and we have witnessed extraordinary growth of developed economies over the 20th century with a concomitant increase in per capita energy usage. However many developed countries have broken the nexus between GDP growth and energy usage and projections based on past usage or expected GDP growth probably overstate potential future emission rates.

One of the factors that has been recognised as being important for future fossil fuel usage is the difference in public infrastructure and access to technology available in different communities. If all countries were to push towards industrialisation and the implementation of modern public infrastructure and services, as is their rightful objective, then emissions rates and atmospheric concentrations of carbon dioxide will increase rapidly. Some national economies, including India and China, have been developing very rapidly in recent years and underscore the potential for expansion of global energy usage. However assumptions that all countries will develop at such rates are considered to be unrealistic by some economists[111], who believe that many countries will remain as low energy economies while the development of others will be tied to greater energy efficiency, both in production and usage.

Estimates of the availability and accessibility of fossil fuel reserves lead to the suggestion that reserves will not be an issue into the 'foreseeable future'. There are two cautionary notes to be made about such stoicism. Firstly, the foreseeable future is a vague concept. Certainly there are sufficient proven accessible reserves to satisfy demand well into the middle of the 21st century, but reserves are becoming more difficult and costly to access. Secondly, some optimistic commentators point to the history of forward estimates that regularly suggested reserves are limited but that more reserves had continued to

---

111 Professor Ian Castles (former Australian Government Statistician) and Dr David Henderson (former head of the Economics Division of the OECD) have criticised the IPCC for advancing global growth scenarios that overstate the rate of growth of anthropogenic emissions. The excess emission rates feed into the computer models for climate projections and are claimed to overstate the potential for global warming.

be proven. To the best of our knowledge, fossil fuels are non-renewable and at some stage they will become exhausted. Combined with the fact that oil and gas are also feedstock to a range of chemical industries there is cause to take a very long view at how and at what rate the global society chooses to burn fossil fuels. However, any decision and agreement to restrict usage must be soundly based policy, not a response to perceived dangers based on inadequate science.

The concept of radiative forcing

Notwithstanding the uncertainties about future emissions and about increased uptake rates by the oceans and the terrestrial biosphere, there is little doubt that anthropogenic emissions will increase atmospheric concentrations of carbon dioxide well into the 21st century. However the arguments advanced by IPCC to support its claims about projected global warming and other climate impacts from increasing concentrations of atmospheric carbon dioxide are fundamentally flawed. Similarly, there is every reason to believe that the variability of global temperature and other climate characteristics that have been experienced over the past century are part of the natural variability of the climate system and are not a consequence of recent anthropogenic activities.

The hypothesis advanced by IPCC in support of its global warming scenario is based on the concept of radiative forcing. The physical basis for the concept is the relative magnitudes of the earth's radiation components that make up the earth's annual and global mean energy budget[112] (see Figure 1). The energy budget is a simplified one-dimensional portrayal of the earth and its atmosphere that gives emphasis to radiation processes at the expense of other processes that make up the complexity of the climate system. The radiative forcing hypothesis assumes that, in an undisturbed state prior to the commencement of industrialisation, the climate system was in equilibrium and the earth was in radiation balance. That is, at the top of the atmosphere[113] reflected solar radiation and emissions of terrestrial radiation balanced incoming solar radiation. The hypothesis is that any change, either to the incoming solar radiation, the earth's albedo[114] or the outgoing terrestrial radiation, will cause an imbalance

---

112 IPCC, 2001. Ibid, Figure 1.2, p90.
113 For practical considerations, IPCC equates the 'top of the atmosphere' with the tropopause. The tropopause is the boundary between the troposphere and the stratosphere. The troposphere makes up about 90 percent of the atmosphere mass and is the layer above the earth's surface where weather systems are active. The overlying stratosphere has stable airflow.
114 The earth's albedo, or reflectance, determines the amount of solar radiation reflected directly back to space. Albedo changes with cloud cover and the nature of the surface (eg, ocean, desert, forests, ice sheets, etc).

to the global energy budget and the climate system will adjust so that a new radiation balance for the earth is established.

Radiation forcing is, therefore, an imposed perturbation on one of the radiation components of the earth's energy budget. If the net radiation received by the earth increases then the earth will warm, and if the net radiation decreases the earth will cool. It is expected that an increased concentration of carbon dioxide in the atmosphere will cause the emission of terrestrial radiation to space to be reduced, and energy will tend to accumulate within the climate system. As a consequence of the accumulating energy, the earth will warm and increase the emission of terrestrial radiation to space until a radiation balance is again achieved at the top of the atmosphere.

The reason that terrestrial radiation to space is reduced when there is an increase in the concentration of carbon dioxide in the atmosphere relates to the prevailing temperature lapse rate (the temperature change with height) of the troposphere. At any layer in the troposphere there is emission of terrestrial radiation both upward and downward, and there is also absorption of energy emitted from layers below and above. Because the emission from a layer is a function of temperature and the temperature of the troposphere decreases with height then each layer emits less radiation than the one below. Overall, the intensity of upward directed terrestrial radiation decreases with increased height above the surface, and the terrestrial emission to space is much less than that emitted by the relatively warm surface of the earth. This is the greenhouse effect.

The emission of terrestrial radiation to space is largely the emission from the uppermost layer where there is sufficient greenhouse gas to be effective in absorbing the radiation from below and for itself to emit radiation. By increasing the concentration of atmospheric carbon dioxide the final emission is pushed to a higher and colder layer, which emits less terrestrial radiation. Calculations demonstrate that a doubling of the concentration of carbon dioxide will reduce the terrestrial radiation from the top of the atmosphere to space by about 4 W/m^2. In order for the intensity of terrestrial radiation to return to previous levels (ie, to balance the solar input) it is necessary for the whole earth-atmosphere system to warm.

Although the upward emission of terrestrial radiation reduces with height, and the emission to space is less than the emission from the surface, this does not mean that the atmosphere is absorbing terrestrial radiation and being warmed. Each layer in the troposphere is also absorbing the radiation emitted from the layer above and is itself emitting more radiation downward because it is warmer than the layer above. If we consider the troposphere as a deep layer then the radiation balance of Figure 1 shows that more terrestrial radiation is emitted to space (from the top of the layer) and to the earth's surface (from the bottom of the layer) than is absorbed from the surface emission[115]. The atmospheric greenhouse gases (water vapour, carbon dioxide, etc), clouds and aerosols cause the troposphere to continually cool, not warm as is often portrayed in the popular press.

The radiative forcing concept is simple and plausible but does not realistically represent the climate system and its processes. Major deficiencies can be summarised as:

- the one-dimensional model, with no latitudinal variation, only has application on a flat earth. The earth is near spherical and receives excess solar radiation over tropical latitudes and outgoing longwave radiation exceeds incoming solar radiation over middle latitudes and polar regions. The atmospheric and ocean circulations that transport energy poleward cannot be ignored if the local net radiation and surface temperatures are to be reconciled. Nowhere is there local radiation balance at the top of the atmosphere and nowhere is the local surface temperature a function of radiation alone. The atmosphere and ocean circulations are crucial for transporting energy, maintaining a near global energy balance and for regulating surface temperatures, especially over polar regions;

- the assumption of net radiation balance at the top of the atmosphere cannot be verified. Estimates of radiation at the top of the atmosphere using satellite observations[116] give a possible error of about 9 W/m^2 in net radiation, mainly due to the limits of sampling imposed by only two orbiting satellites. We do not know if there is now, or has recently been, net radiation balance at the top of the atmosphere. The uncertainty level is greater than the IPCC estimate for anthropogenic radiative forcing (currently 2.43 W/m^2 for anthropogenic contributions to the atmospheric concentrations of well-mixed greenhouse gases);

115 The troposphere layer absorbs 67 W/m^2 of solar radiation and 390 W/m^2 of terrestrial radiation emitted from the earth's surface, a total of 457 W/m^2. The layer emits 235 W/m^2 to space and 324 W/m^2 downward to the surface, a total loss of 559 W/m^2. The net radiation loss of the troposphere layer, or radiation cooling, is 102 W/m^2.
116 Trenberth, K.E and A. Solomon, 1994. Ibid

- the simple radiative forcing model takes no account of the energy reservoirs of the climate system. In particular, the surface layers of the oceans directly absorb a large fraction of incoming solar radiation. The rate of transfer of heat and latent energy from the ocean to the atmosphere is a function of surface temperature and wind. Also, the mountain glaciers and polar ice sheets represent a net transfer of energy from the oceans; they expand and contract with variations of the meridional energy transport;

- the one-dimensional energy budget model suggests that evapotranspiration and thermals are the mechanism for mixing the excess solar and net longwave radiation from the earth's surface into the troposphere to offset net longwave radiation cooling. This simplistic explanation ignores the important role of deep tropical convection and atmospheric overturning for regulating the temperature of the tropical troposphere and the rate of heat transport through the climate system.

The one-dimensional energy budget model is a construct that has successfully been used in the past to provide computational closure when estimating the magnitude of radiative components of the climate system. The value of the model is that, for the purpose of computing the global energy budget, the net effects of dynamic processes of the atmosphere and ocean are averaged to zero. With the assumptions of 1) no net exchange of energy between the atmosphere and the underlying ocean and land surfaces, and 2) net radiation balance at the top of the atmosphere, then globally averaged annual mean radiation components can be calculated with energy exchanges across the earth-atmosphere interface as residuals. Errors resulting from the assumptions contribute to the errors of the estimated radiation values. Also, it should not be interpreted that the dynamic processes have no importance in shaping the climate system. The model is a far cry from how the climate system actually works.

An example of how the one-dimensional radiation forcing model has led to erroneous, even bizarre, conclusions is the derivation of a supposed surface temperature response to radiation forcing. IPCC used the model to investigate "the equilibrated global, annual mean surface temperature responses to radiative perturbations caused by changes in the concentrations of radiatively active species" [117]. The application is claimed to lead to a near invariant temperature response to radiative forcing (a climate sensitivity parameter) of about $0.5°C/Wm^{-2}$.

---

117 IPCC, 2001 Ibid, Appendix 6.1, p405.

·r such a response is erroneous and misleading.

IPCC temperature response relationship can be derived explicitly from an inversion of the well-known Stefan-Boltzmann Law that relates the intensity of radiation emission to the temperature of the radiating surface[118]. Such an inversion can only be true if the underlying surface has no thermal capacity, there is no conduction or evaporation, and the surface temperature is in equilibrium with the radiation source. However the derived sensitivity of $0.18°K/Wm^{-2}$ is only one third the value obtained by IPCC.

For the earth, the correct application of the Stefan-Boltzmann Law is in calculating the intensity of terrestrial radiation emitted from the earth's surface at a given temperature. The intensity of radiation varies explicitly with the fourth power of surface temperature. However the inverse relationship does not hold and the inversion of the Stefan-Boltzmann Law to calculate a temperature response to changes in radiative forcing is a complete nonsense.

Surface temperature is calculated from an energy budget that involves upward and downward terrestrial radiation, absorption of solar radiation, eddy exchanges of heat and latent energy with the overlying atmosphere, and upward or downward conduction of heat in the ocean or land surface layer. Many of these factors, including emission of terrestrial radiation and eddy exchanges of heat and latent energy, are a function of the surface temperature and cannot be ignored. Indeed, at surface temperatures typical of the tropical oceans it is latent energy that is the dominant form of energy loss from the surface because terrestrial radiation emissions are nearly balanced by downward radiation from the warm and moist atmospheric boundary layer. Over the tropical oceans the latent energy exchange moderates the temperature response to all forms of radiation forcing, including solar radiation. Daytime ocean surface temperatures are cooler than the corresponding land temperatures at similar latitudes, especially the dry surfaces of deserts.

IPCC also concludes, from experiments carried out with computer models representing the three-dimensional coupled atmospheric and ocean circulations, that radiative forcing is also a good estimator for the global mean surface temperature response. However, IPCC adds the qualifier, "but not to a quantitatively rigorous extent as in the case of

---

118 Stefan-Boltzmann Law: $F = \sigma T^4$, where F is radiation intensity and T is absolute temperature. An inversion of the equation leads to the rate of change of temperature with changes in radiation. That is, the climate sensitivity parameter claimed by IPCC is expressed by the relationship $dT/dF = T/4*F$. For values typical of the earth's surface temperature $(278°K)$ and emission intensity $(390\ W/m^2)$ the sensitivity parameter is only about 0.18 and one third the value claimed by IPCC.

one-dimensional radiative-convective models". The interpretation that there is a linear response of the non-linear climate system is surprising. In addition, the range of responses to similar radiative forcing by different computer models is indicative of the uncertainty of formulation and specification of boundary processes in the various computer models.

An additional limitation of the one-dimensional radiation forcing model is its inability to recognise and take into account the energy transfers associated with expansion of the polar ice sheets and mountain glaciers during long-term snow accumulation. The transfer of water from the oceans to snowfields and ice sheets represents a systematic transfer of energy from the oceans. Latent energy is extracted from the oceans during the formation of water vapour but the latent energy released to the atmosphere during snow formation is greater than the latent heat taken up from the oceans during evaporation. Overall, there is a net transfer of energy from the oceans to the atmosphere during that part of the hydrological cycle when snow accumulates. This is also the case during sea ice formation.

The net radiation excess over the tropics and the deficit over polar regions is a difference that is not accommodated within the one-dimensional construct. Global energy balance can only be maintained through transport of energy from the tropics to polar regions by the atmospheric and ocean circulations. We have also noted that the atmospheric transport can either be as potential energy (meridional overturning), as heat (Rossby Waves and weather systems) or as latent energy (also Rossby Waves and weather systems). At this time there is only rudimentary understanding as to why the thermodynamics of the atmosphere partitions the transport between the different modes.

Meridional overturning is linked to the Hadley Cells of the tropics and the condensation and latent heat release occur largely in the warm equatorial region. During meridional overturning there is no net loss of energy from the ocean because evaporated water returns as tropical rainfall. However, during the transport by Rossby Waves and weather systems there is significant precipitation as snowfall, particularly over subpolar and polar regions. During snow formation and accumulation there is a net loss of energy from the oceans linked to the mass of water transferred to the polar ice sheets. Internal variability of the climate system that changes the rate of meridional transport of energy, and/or the partitioning of energy transport between meridional overturning and Rossby Waves and weather systems, will affect the rate of accumulation of snow and energy loss from the oceans.

One of the basic assumptions of the IPCC construct of the one-dimensional radiative-forcing model is that a stable climate is linked to balanced radiation at the top of the atmosphere. It is informative to carry out some simple calculations to compare the projected anthropogenic greenhouse radiative forcing with radiative forcing during a glacial cycle, a period of extraordinary climate extremes.

It has already been noted that climate characteristics changed appreciably on the approximately 100,000-year interval associated with the last major ice age. During the cooling period, when polar ice sheets were expanding and thickening, we would expect the earth to be losing energy to space. This net radiation loss is offset within the climate system by the latent energy released from the oceans during snow formation and ice accumulation. Similarly, during the warming phase, as the polar ice sheets melt and contract, a surplus of solar radiation input is required to provide latent energy for the formation of melt-water. There is also a net loss or gain of radiation associated with the respective cooling or warming of the ocean surface layers during the glacial cycle but this is at least two orders of magnitude less than the latent energy exchanges and can be ignored.

We can use the fall and rise of sea level to estimate the mass change of the polar ice sheets during their growth and contraction over the glacial cycle. Oceans cover about 70 percent of the earth's surface and sea level fell and rose more than 130 metres during the most recent glacial event. This estimate of sea level fall suggests about 4.6×10^{16} m^3 of water became locked in the polar ice sheets. The ice sheets formed from falling snow and about 13.2×10^{25} joules of latent energy were released to the atmosphere during the period of expansion and thickening. In total, the latent energy released during expansion of the polar ice sheets represents a global average net loss of radiation at the top of the atmosphere of about 0.1 W/m^2 persisting for about 80,000 years. Similarly, the heat absorbed by the climate system during the melting of the ice sheets is of the order of 1.6×10^{25} joules. The retreat of the ice sheets could be achieved by an average global net radiation input of about 0.1 W/m^2 persisting for about 10,000 years.

The estimated global net radiation imbalances at the top of the atmosphere during the glacial cycle are, on average, very small[119]. They are significantly less than the radiative forcing of about 4 W/m^2 estimated for a doubling of greenhouse gas concentrations. However, before developing scenarios on the potential climate impact of anthropogenic increases in atmospheric greenhouse gas concentrations drawn from the glacial events we must be very clear about the different global energy constructs.

---

119 These simple calculations assume a steady advance and retreat during the overall glacial cycle. They do not take account of the more rapid fluctuations within the envelope of the cycle that would require significantly greater top of the atmosphere global net radiation imbalances during these higher frequency fluctuations.

During a glacial cycle there is a sustained seasonal cycle of solar radiation that is forced and maintained by external factors (eccentricity of the earth's orbit). The seasonal cycle of solar radiation imparts a modulation to the energy input over the tropics and to the rate of meridional energy transport by the climate system. A sustained reduction in poleward energy transport when the earth's orbit is eccentric will lead to polar cooling and the advance of the ice sheets, while an increase in the poleward energy transport as the orbit becomes more circular will lead to polar warming and retreat of the ice sheets.

In the case of increasing atmospheric greenhouse gas concentrations the forcing is brought about by internal processes that initially lead to a relatively uniform reduction in the global longwave radiation to space. The IPCC refers to the adjustment of global surface temperature that results from an incremental increase in net global radiation as climate sensitivity. The sensitivity is based on radiation balance in a simple hypothetical model. In reality, climate sensitivity is complex and involves the effective thermal capacity of the climate system, particularly the energy reservoir of the ocean surface mixed layer, and processes regulating the rate of meridional energy transport[120]. Over the tropics, evaporation and latent energy processes regulate surface temperature change. Over polar regions the surface temperature is warmer than might be expected from radiation balance because of the import of energy from the tropics.

Surface temperature and the potential for snow accumulation over polar regions are particularly sensitive to the rate at which energy is imported from the tropics. A reduction in the meridional transport of energy by the atmospheric circulation leads to cooler surface temperatures and a larger area with temperatures persisting below 0°C. This latter greatly assists winter snow to persist through summer and lead to a long-term expansion of the snowpack. Conversely, an increase in the meridional transport of energy by the atmospheric circulation will lead to warming of polar regions and melting of the snowpack. It is instructive to note that, for the northern hemisphere, a sustained one percent increase for 7 years in the average meridional transport of energy by the atmospheric circulation will transport sufficient energy to melt the existing Arctic Ocean sea ice.

The known variability of the climate system is not adequately explained by a one-dimensional construct such as the radiation

---

120 The ocean surface layer alone has a thermal capacity approximately two orders of magnitude (100 times) greater than the overlying atmosphere and provides thermal inertia to the climate system. Early computer models represented the oceans as uniform slabs with prescribed heat exchange properties.

forcing hypothesis used by IPCC. Such a construct cannot accommodate the ramifications of latent energy exchange as polar ice sheets expand and contract. In its application by IPCC there is no recognition of the role of the oceans as an energy reservoir for the climate system nor of the thermal capacity of the oceans that diminishes the impact of radiation forcing. The proposal for a climate sensitivity that relates global surface temperature change to radiation forcing is simplistic and implausible because of the complex response of tropical oceans (dominated by evaporation) and the reliance on polar regions for import of energy from the tropics to maintain current surface temperatures.

Solar Influences on Climate

Solar radiation is the energy source that not only maintains the warmth of the earth but also energises the atmospheric and ocean circulations. It is not surprising, therefore, that the sun and the strength of solar radiation (the solar irradiance) have been studied in great detail. It has been only since the availability of satellites that direct measurements of solar irradiance outside of the earth's atmosphere have been possible. Prior to satellites, any measurement of the solar beam has had to contend with the effects of the atmosphere and its modification of radiation. Even balloon-borne instruments could not reach sufficient altitude to eliminate all effects of the upper atmosphere.

Despite the increasing precision of instruments developed during the 19th and 20th centuries the variance in the measurements was relatively large due to scattering of solar radiation by aerosols and cloud particles, and by the selected absorption by greenhouse gases and atmospheric particles. The variations in the measurements could not be attributed to variability or trends in solar irradiance with any certainty. For practical purposes, a universally agreed value of solar irradiance was treated as the Solar Constant. However, it was suspected for a long time that solar irradiance is variable. With the development of satellite-borne instruments, it has been possible over the past two decades to monitor the sun's irradiance outside the atmosphere's influence and estimate more accurately the magnitude of solar variability.

In spite of their better observing position the satellite instruments have their own set of problems. The instruments have great precision and are able to detect the magnitude of significant short period fluctuations that previously were not recognised. However each instrument only has a working life of a few years and sensors tend to deteriorate with age. The most significant problem has been the lack of absolute calibration. Different instruments varied by up to 7 W/m2 in their nominal recorded values. To an extent these problems have been eliminated by careful comparison between overlapping records of different instruments but the absolute calibration and modern solar irradiance has not been fully resolved.

The confirmation that solar irradiance is variable has been the basis for attempts to reconstruct values from earlier times. Sunspot numbers have been found to vary with changing solar radiation output from the sun and are one proxy used for the reconstruction of the record of solar irradiance. The development of telescopes early in the 17th century made it possible to observe the sun in some detail, including the occurrence of sunspots. Records of the number of sunspots have been kept since the middle 17th century and methods have been developed to link changing sun spot numbers with intensity of solar irradiance. The long record of changing sunspot numbers has provided a basis for estimating the change in magnitude of solar irradiance over this time[121], although different methods give differing reconstructions.

The reconstructions of solar irradiance based on sunspot numbers suggest that values have varied between about 1364 W/m^2 and 1366 W/m^2 until the beginning of the 20th century. Lowest values occurred during the late 17th century (the long Maunder Minimum) and the early 19th century (the shorter Dalton Minimum). During the later part of the 19th century solar irradiance is estimated to have weakened to about 1365.5 W/m^2 before a period of sustained increase to values of about 1367.5 W/m^2 by the end of the 20th century.

Several other proxies have been identified as potentially suitable for reconstructing total solar irradiance from earlier times. One method[122] is to use concomitant variations in production of the isotopes of carbon (^{14}C) and beryllium (^{10}Be) as a proxy for total solar irradiance. This reconstruction suggests that there was a peak of solar irradiance in the 12th century that had similar intensity to current values. Intensity declined to a minimum in the 15th and 16th centuries before generally recovering, with breaks for the Maunder and Dalton minima, to current values. This pattern is shown in the pattern of isotopic carbon (^{14}C) shown in Figure 25.

Many of the semi-quantitative patterns of solar irradiance for the past millennium that have been reconstructed from proxies have similarities to the global pattern of advance and retreat of many mountain glaciers. Minimum irradiance values after the 12th century correspond to the advance of mountain glaciers during the Little Ice

121 Lean, J., J. Beer and R. Bradley, 1995. Reconstruction of solar irradiance since 1610: implications for climate change. Geophys. Res. Lett., vol 22, pp3195–3198.
122 Bard, E., G. Raisbeck, F. Yiou, and J. Jouzel, 2000. Solar irradiance during the last 1200 years based on cosmogenic nuclides. Tellus 52B, pp985–992.

Age, and the relatively steady increase since the Dalton Minimum corresponds to the steady glacial retreat and global warming of the 20th century. The similarities, while tantalising, cannot be taken as incontrovertible evidence of solar forcing of the climate system.

A criticism of the solar irradiance reconstructions that has to be addressed is whether the magnitudes of the changes in solar irradiance proposed would input sufficient energy to the climate system to be noticeable. The first issue is to establish the magnitude of additional solar energy intercepted since 1700 (the Maunder Minimum and depth of the Little Ice Age) and whether it would be sufficient to affect climate. The average increase in solar irradiance is taken to be 1.5 W/m^2 over the period 1700–1900 and 2.5 W/m^2 over the 20th century. The total increase in energy interception (assuming a 0.7 albedo for reflected radiation) by the climate system is 1.1×10^{22} joules. If directed entirely to melting land ice the increased energy would yield 3.4×10^{19} cm^3 and raise sea level about 10 cm. For purposes of comparison only, this ice mass is equivalent to the volume of sea ice in the Arctic Ocean.

The above order of magnitude calculation supports speculation that there could be a causative link between solar irradiance increase since the late 17th century Maunder Minimum and global warming since the Little Ice Age but there are outstanding issues. Most of the solar irradiance increase would be absorbed by the tropical oceans and not over polar regions. However, as with the bulk of solar radiation reaching the earth, the additional energy would follow a path from the tropical oceans to the atmospheric boundary layer; via deep convection to the troposphere; and by the meridional circulation, Rossby Waves and weather systems to higher latitudes. The transport path means that

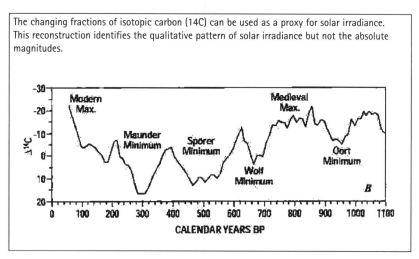

The changing fractions of isotopic carbon (14C) can be used as a proxy for solar irradiance. This reconstruction identifies the qualitative pattern of solar irradiance but not the absolute magnitudes.

Figure 24: The isotopic carbon (^{14}C) record over the past 1100 years

additional solar energy is available to warm the earth's surface and melt polar ice. A critical point is that the additional solar insolation is retained in the climate system until the earth's surface over middle latitudes and polar regions is warmed. This last factor ensures that the earth's climate response is sensitive to variations in solar irradiance.

In its discussion on varying solar radiation, the IPCC comes to a different conclusion about the overall importance of changes in solar irradiance. IPCC considers changes in solar irradiance as another form of radiative forcing that can potentially act on the climate system. That is, the solar forcing is the change in solar irradiance from an arbitrary period of 'climate balance'. IPCC takes 1750 as the date of commencement of industrialisation and benchmark for solar forcing. In its application, IPCC takes solar forcing to be the difference between contemporary solar irradiance and that of 1750, as modified by a geometric factor to convert irradiance to a global average forcing.

There are two major issues with the IPCC approach, one acknowledged by the IPCC and one not. Firstly, IPCC recognises that its methodology for solar forcing is sensitive to the reference date. A choice of 1700 (the end of the Maunder Minimum) would give a value of solar forcing twice as large as the value for 1750; a choice of 1776 would have given a significantly smaller value. Secondly, the geometric factor used by IPCC to convert irradiance to radiative forcing is again an application of flat-earth physics. The reduction of the solar irradiance by the earth's albedo is valid because it recognises that part is reflected back to space without interacting with the climate system. However, IPCC also reduces the solar irradiance by a factor of four on the basis that the total solar radiation intercepted across the area of the earth's disc is spread over all the area of the globe. The area of a sphere is four times the area of a disc of the same diameter. This latter does not acknowledge that the bulk of solar insolation is received in the tropics. The method overestimates the change to solar insolation received over the polar regions and underestimates that received over the tropics. The IPCC underestimates the energy input into that part of the globe from where the climate system is energised.

An impediment to causatively linking changing solar irradiance with global surface temperature variations is the apparent diverging of the trends of solar irradiance and surface temperature over the past two decades when observational accuracy is at its best. The warming trend of global surface temperature between 1980 and 2000 is about 0.3°C, slightly less than the 0.4°C warming from 1910 through 1940[123]. The reconstructions of the total solar irradiance for the first half of the 20th century give increases ranging between 2 W/m^2 and 4 W/m^2. For

123 WMO, 2001. WMO statement on the status of the global climate 2000. WMO No. 920, Geneva.

the recent warming period, when satellite observations cover two solar cycles, there apparently is little increase. However, the issue is not fully resolved because of uncertainty introduced in the calibration of data obtained from instruments with differing characteristics from overlapping satellite records. A recent reconstruction of the satellite data does suggest that a continuing upward trend in solar irradiance is a more likely interpretation[124].

Variations in the annual cycle and geographic distribution of solar insolation across the earth's surface also occur because of the periodic changing of the earth's orbital parameters. Each parameter affects the distribution of solar insolation over the earth by latitude and season in a different way and the signals of the 22,000-year precession, the 40,000-year obliquity and the 100,000-year eccentricity periods can be found in many very long palaeoclimate records. The changing orbital parameters do not appreciably change the total solar insolation absorbed by the earth at any time and controversy surrounds the mechanisms by which the various phases are able to stimulate particular climate responses, especially the ice age epochs associated primarily with the 100,000-year eccentricity period. The signals in the palaeoclimate records suggest a sensitivity of the earth's climate to solar variations even though the processes have yet to be identified.

The variations in temperature and waxing and waning of polar ice sheets during glacial and interglacial epochs have, in the past, been treated as a local radiation problem. The focus of investigation has largely been on the variation of solar insolation over high latitudes, especially at the margins of winter snow cover and permanent ice. The inclusion of meridional transport of energy and of latent energy exchanges associated with the hydrological cycle into the analysis opens up new potential avenues for climate response.

When the earth is in near circular orbit around the sun, as it is now, there is very little annual variation of solar insolation over the tropics. Meridional transport of energy between the tropics and the polar regions is mainly regulated by the annual cycle of radiation deficit over the respective polar regions. The annual temperature cycle is also regular and a balance is reached between winter snow accumulation and summer snow melt. A regular annual cycle of snow cover in each hemisphere ensures that the net amount of latent energy extracted from the climate system during winter snow accumulation is returned during summer when snow melts.

The meridional transport of energy between the tropics and the polar regions becomes irregular as the earth's orbit becomes more elliptical. Although the total annual insolation over the tropics is little

124 Willson, R.C. and A.V. Mordvinov, 2003. Secular total solar irradiance trend during solar cycles 21–23. Geophys. Res. Lett., vol 30, N0. 5

changed between the circular and elliptical orbits, as the orbit becomes elliptical an annual cycle of solar insolation is introduced over the tropics. During the perihelion sector of the orbit, solar insolation is increased over that for a circular orbit; during the aphelion sector, solar insolation is reduced. Also, because of the orbital geometry, the time of increased solar insolation during the perihelion is reduced and the time of reduced insolation during the aphelion sector is increased.

The calculations that point to very small global radiation imbalances during the glacial epochs and the intervening interglacials imply that a persisting forcing is the main requirement for regulating the climate system over such long intervals. The period over which orbital eccentricity changes is a good match for the duration of the combined glacial and interglacial epochs. However the details of how seasonally changing solar insolation over the tropics and meridional transport of energy combine with the hydrological cycle to bring about persisting loss and gain of global radiation over the epochs are yet to be fully elucidated.

Variations in the cosmic ray flux reaching earth has been proposed as a quite different mechanism linking solar activity and climate response. Cosmic ray flux affects ionisation in the atmosphere and the rate of production of charged aerosols. Atmospheric aerosols provide a nucleus on which cloud droplets form and grow and the formation and growth of cloud droplets is favoured when the atmospheric aerosols have an electrostatic charge. The low-level cloudiness of the earth is expected to increase when there are more charged aerosols present. When there are more low clouds present the average albedo of the earth will also increase and a larger proportion of solar radiation will be reflected directly back to space. Therefore, an increased cosmic ray flux and its tendency to increase the electrostatic charge on aerosols is expected to lead to more low-level clouds and greater reflection of solar energy to space.

During a period of increased cosmic ray flux it is expected that there will be more low level clouds, a higher global average albedo and more reflection of solar radiation to space. There will also be less solar energy absorbed in the climate system to warm the earth. The theoretical linkage relates a brighter sun to an enhanced solar wind, which diminishes the cosmic ray flux and, through reduced low-level clouds, results in more absorbed solar energy and warming of the earth. Satellite observations between 1985 and 2000 indicate that reflected solar radiation to space was reduced during the period, with most of the reduction in the 1990s[125]. However the reduction in albedo is not attributed to an increase in cosmic ray flux but to an increase in the tropical Hadley Cell circulations and increased subsidence resulting in fewer clouds.

125 Chen, C., B.E. Carlson and A.D. Del Genio, 2002. Evidence for strengthening of the tropical general circulation in the 1990s. Science, vol 295, pp838–841.

Support for the linkage between cosmic ray flux and climate comes from a reconstruction of global temperatures over the past 500 million years based on oxygen isotope histories recovered from calcitic shells in sedimentary deposits[126]. The cold periods of this temperature history are found to match the times of extreme latitudinal extension of ice rafted debris over the ocean floor of the North Atlantic Ocean, identified as periods of "icehouse" conditions on earth when glaciation lasted millions of years. The average period between the very cold "icehouse" and intervening warm "greenhouse" conditions is found to be about 135 million years and, within the limits of dating uncertainty, a close match to the approximately 143 million-year interval between the solar system passing through the spiral arms of the galaxy. There is a significant increase in galactic cosmic ray flux when the solar system passes through the spiral arms of the galaxy because most star formation activity is within the spiral arms. The increase in galactic cosmic flux within the spiral arms is much larger than variations produced by fluctuations of solar wind. On this basis, the recent 100,000-year glacial cycles are within the current "icehouse" envelope that commenced a few tens of millions of years ago.

Needless to say, the cosmic ray flux theory of climate variability is controversial. The limited data set of approximately 4,500 oxygen isotope records necessarily ensures that the temperature record lacks structural detail. It is claimed, however, that the data do resolve four "icehouse" periods over the last 500 million years. Reconstruction of the cosmic ray forcing is dependent on modelling, taking account of the known geometry and dynamics of the spiral arms of the galaxy and assumptions about cosmic ray generation in the spiral arms. Application of the theory to recent climate history, including the past millennium, is limited by the lack of analogous cosmic ray flux data.

The patterns of climate variability on various timescales that can be identified in the palaeoclimate record give a level of credence to the possibility of climate forcing being a reflection of celestial origins. For nearly a century the Milankovitch theory for explaining major glacial events, lasting approximately 100,000 years, has held sway. This is despite a limited understanding of how changes in the eccentricity of the earth's orbit around the sun can cause waxing and waning of the polar ice sheets, especially those of the northern hemisphere. Similarly, the prolonged humid period of North Africa during the early Holocene is linked to a period of maximum solar insolation during the northern hemisphere summer that is a consequence of the 20,000 year precession of the earth's axis of rotation. However, the abrupt change

---

126 Shaviv, J. and J. Veizer, 2003. Celestial driver of phanerozoic climate? GSA Today, July pp 4–10.

from humid to dry conditions about 5,000 years ago[127], and many other relatively abrupt climate changes are not consistent with slowly changing insolation. The abrupt changes suggest other, as yet unidentified, climate processes are also important.

There is also speculation that the earth's climate system does not respond linearly to changes in solar insolation, suggesting that there may be several (or even many) relatively stable climate states. Moreover, the relatively abrupt shift from one state to another opens the possibility that a small change in external forcing can trigger major readjustments to internal relativities between energy reservoirs, transports and sinks. Critical internal factors may include the accumulated energy in the tropical ocean reservoir, the magnitude of the polar ice sheets that buffer a net transfer of energy and mass from the oceans, or the strength of the ocean thermohaline circulation and its interactions with the atmosphere.

There is not enough quantitative data relating to changing solar irradiance and cosmic ray flux to be definitive about their effects on climate and recent global temperature variations. However there is sufficient qualitative information to suggest that it would be unwise to discount solar factors as being a significant contributor to recent climate variability, especially that of the past millennium leading to the recent warming of surface temperatures. Sunspot numbers since the 17th century, and the isotopes of carbon and beryllium extending over the past 1,000 years, show qualitative agreement with known global advance and retreat of mountain glaciers. The IPCC relegation of solar forcing to a minor role is based on a flawed theoretical framework that underestimates the potential solar contribution to climate variability and is against a body of qualitative evidence.

---

127 Thompson, L.E. et al, 2002. Ibid.

Internal Variability of the Climate System

In its Third Assessment Report the IPCC has concluded that there is new and stronger evidence that "most of the warming (of the global surface temperature) observed over the last 50 years is attributable to human activities". The claim is based on an expressed belief that progress has been made in reducing uncertainty about the responses of the climate system to different external influences. The claimed major areas of progress relate to understanding how climate has actually varied over the recent past, and in the development of computer models that better quantify the relative contributions of natural and anthropogenic factors to climate change. The IPCC confidence that global warming is a consequence of anthropogenic causes is strongly linked to its conclusion that the warming of the past 100 years "is very unlikely to be due to internal variability alone, as estimated by current models".

The reconstruction of global surface temperature over the past 1,000 years (see Figure 7), especially the limited variability of the pre-instrument period, is the basis for the IPCC claim of better understanding of how climate has actually varied over the recent past. IPCC chooses to give weight to the reconstruction despite the fact that the envelope of uncertainty is as large as the range of variability being investigated. Ignoring the uncertainty envelope, the IPCC argues that because there is only limited global surface temperature variability until the 20th century then the warming of the 20th century must be unusual. In addition to the low skill level, the millennium reconstruction of northern hemisphere surface temperature used by IPCC has already been demonstrated to be grossly deficient in its methodology and execution and is misleading[128].

---

128 See the earlier discussion on "Statistical methods for climate reconstruction".

It is beyond belief that IPCC accepts that only 12 proxies are sufficient to reconstruct northern hemisphere surface temperatures for the period 1000 to 1400, and that these should be thought to truly represent a pattern of hemispheric variability. Not only is each proxy a very imprecise measurement of the annual average surface temperature for a limited region of the globe but 12 proxies are insufficient independent records to give a fair representation of the northern hemisphere average. This deficiency is compounded by the fact that three of the proxies are component parts of the same regional tree ring data set and four others represent sites in the southern hemisphere! These deficiencies in the methodology and its ability to reconstruct northern hemisphere temperature stand beside the published claim that there were serious compilation and data management errors in the mechanics of the reconstruction[129].

The pattern of the reconstruction is also very difficult to accept when assessed in the context of the wide body of cultural and historical evidence from Europe and the North Atlantic Ocean regions. The historical evidence clearly identifies a Medieval Warm Period followed by a Little Ice Age. Scandinavian settlements established in Iceland and Greenland benefited from a relatively warm and long lasting period when northern seas were more benign and ice-free. This expansion of settlements is well documented. The Greenland settlements relied on trade for their survival and it was only when navigation became too hazardous, even for ships more seaworthy than their predecessors, that the settlements were literally abandoned. The onset of colder conditions across northern Europe in the 14th century is also well documented, for example with advancing glaciers causing the abandonment of farms in Norway and mining activities in the high Alps of Switzerland. The advance and retreat of glaciers over later centuries has been documented as a worldwide pattern.

Knowledge of meteorology and meteorological processes also leads us to the conclusion that IPCC's claim of the Medieval Warm Period and Little Ice Age being phenomena confined to the northern European and North Atlantic regions is also a nonsense. Rossby Waves and weather systems over the middle and high latitudes transport energy towards the poles. Occasionally the Rossby Waves become locked into stationary patterns, known as blocking patterns, for many months. During a blocking pattern, a particular region will persistently come under the influence of cold polar winds while the downstream region comes under the influence of the returning warm wind from the direction of the equator. The blocking pattern also affects regional rainfall patterns. Persistent blocking is one of the reasons why a region will experience unusually warm or cold seasonal conditions. If, as

---

129 McIntyre, S. and R. McKitrick, 2003, Ibid.

IPCC infers, blocking patterns were organised over Europe and the North Atlantic to give about 400 years of unusually warm weather followed by 400 years of unusually cold weather then this would itself represent a major (although most unlikely) climatic fluctuation of global significance.

There is a more realistic explanation for the regional temperature pattern of prolonged warmth, then prolonged cold, and then the more recent return to warm conditions. The explanation relates to variations in the poleward transport of energy by the atmospheric circulation. The Northern Europe-North Atlantic region is within that part of the northern hemisphere where there is net annual radiation deficit. Temperatures over the higher latitudes and polar region are only maintained by import of energy from the tropics. Temperatures will warm when there is a persistent enhancement of the energy transport from the tropics but the temperatures will cool if the meridional transport of energy weakens. Although the path of the energy transport may be temporarily enhanced over particular longitudes, as happens during El Niño events, the transport is on an hemispheric scale and regulated by the Rossby Waves and weather systems embedded in the middle latitude westerly wind flow. Any strengthening or weakening of the poleward transport of energy by the atmospheric circulation will not be confined to a limited geographical region but will effect all of the middle and high latitudes. The effect of the changing atmospheric circulation will also be felt in both hemispheres.

The Medieval Warm Period and the Little Ice Age that are so well documented for Europe and the North Atlantic are almost certainly to have been part of a global waxing and waning of the climate system. The intensity of impact is likely to have varied regionally because of the preferred seasonal patterns of the Rossby Waves and there may have also been some locally unusual effects. But the IPCC contention, that surface temperatures over the northern hemisphere had little variability for 900 years before the introduction of instrument observations, really beggars belief.

In addition to the claimed limited variability of global surface temperatures for the first 900 years of the past millennium, IPCC claims that the warming of the past 100 years is very unlikely to be due to internal variability of the climate system alone. This latter claim is based on computer model studies. However, the climate system is essentially two separate fluids interacting on a rotating sphere and all subject to tidal forcing and solar heating that varies with latitude and season. We should expect interaction between the oceans and the atmosphere to produce variability in their respective circulations. It is the variability of the atmospheric circulation, forced by changing sea surface temperatures, that is the reason attributed for much of the observed interannual to decadal variability of local climate. Scientists are yet to resolve the amplitudes and timescales of variability beyond the interannual timescales of ENSO. It is foolish to dismiss internal variability on these longer timescales until the subject has been thoroughly researched.

Overall, the evidence presented by IPCC in support of its claim of a relatively benign climate system is unconvincing. This view is underscored by IPCC's own findings that there is still considerable uncertainty in the magnitude of internal climate variability[130]. The IPCC compared the variability characteristics of global mean temperature from a selection of computer models with observations. However, the short record of global surface temperature by instruments means that the comparisons are limited to recurrence periods of up to a few decades. IPCC offers no evidence, other than its computer model simulations, that the climate system has neither natural variability nor internal variability on centennial to millennial timescales. Even with the limitation of only considering variability out to multi-decadal timescales, four of the nine models assessed are acknowledged by IPCC to have serious limitations in their representation of variability on those timescales. Many of the models that were compared even had deficiencies at the interannual to multi-year timescale associated with ENSO.

It is very heroic for IPCC to conclude, based on 'model studies', that warming over the past century cannot be explained by internal variability. The computer models generally underestimate internal variability on all measurable scales. In addition, the construction of computer models limits their utility for being a reliable guide to the internal variability of the climate system. Computer models are designed so that, without external forcing, they reproduce a stable climate, notwithstanding that the prototype (earth) may be experiencing natural climate change over a range of timescales.

In an honest assessment within the Third Assessment Report the scientists acknowledge:

> *"There are still severe limitations in the ability of models [computer models of the coupled atmosphere and oceans] to represent the full complexity of observed variability and the conclusions drawn here about changes in variability must be viewed in the light of these shortcomings (Chapter 8)."[131]*

---

130 IPCC, 2001. Ibid. p705.
131 IPCC, 2001. Ibid. p570.

The confidence expressed by IPCC in its Summary for Policymakers (the agreed intergovernmental text) is clearly misplaced. In reality, it is not possible to confidently ascribe recent warming to either internal variability of the climate system, natural forcing (for example, by solar variability), human activities, or a combination of these.

The concept of a stable climate system with global energy balance at the top of the atmosphere, as portrayed by the IPCC, is a simplification that leads to erroneous conclusions. This conceptual framework ignores the very real imbalances of net radiation at the top of the atmosphere, both with latitude and season. The framework also ignores the vital roles of the atmosphere and oceans for transporting energy from regions of net energy excess to those of net energy deficit. Moreover, there are small-scale processes that regulate the internal exchange of heat between the oceans, the atmosphere, the land surfaces and the polar and mountain ice sheets. In the one-dimensional view of the climate system that pervades the IPCC framework, these primary components of the climate system are dismissed as of no importance.

Even without natural or anthropogenic forcing we should expect considerable internal variability of the climate system. The oceans and the atmosphere are both fluids whose circulations interact by way of heat, moisture and momentum exchanges across the ocean-atmosphere interface. The oceans and the atmosphere also interact with the solid earth, especially through the exchange of momentum resulting from the action of surface wind stresses. Interactions between the oceans and the atmosphere are continually stimulated by the seasonally changing pattern of solar radiation and are expected to generate internal variability of the climate system on a wide spectrum of timescales.

Before looking in detail at processes within the climate system that contribute to the generation of internal variability, it is appropriate to summarise our knowledge of the timescales over which the climate has fluctuated before the impact of human activities. We have noted the evidence for enormous fluctuations on geological timescales as the climate varied from 'icehouse' to 'hothouse' conditions over intervals of more than 100 million timescales. The polar regions have been consistently frozen over the past several million years, with the ice sheets expanding and contracting on approximately 100,000-year cycles. From calculations of latent energy exchange, as the ice sheets fluctuate between glacial and interglacial epochs, it is clear that very small but persisting radiation imbalances are quite adequate to accommodate the identified advance and retreat of polar ice sheets. A near global energy balance does not preclude the possibility that a slow but ultimately momentous change is taking place in the characteristics of the climate system. This further underscores the need to consider the importance of internal climate processes.

Within the glacial epochs there is abundant evidence that there were recurring climate fluctuations on the millennial scale that had a significant impact on the extent of glaciation. The cause of these changes remains a mystery. Ice core analyses suggest major changes of temperature have taken place over just a few decades. Layers of ice-raft debris identified in North Atlantic ocean-floor sediment cores confirm recurring periods of abrupt change as being characteristic of the glacial epoch, at least over the North Atlantic Ocean. These shorter timescale fluctuations would be impossible without large changes in the magnitude of energy stored in the ocean and ice sheet reservoirs. There is scant evidence that variations in solar irradiance were forcing these climate fluctuations but, because of lack of relevant data, the possibility cannot be excluded.

Various proxy data relating to regional climate of the past 8,500 years suggest that the Holocene was a relatively stable and benign period in comparison to the preceding glacial epoch. With the exception of the Younger Dryas event as the climate was emerging from the last glacial period, when the climate suddenly reverted to very cold conditions, the climate system has not shown the abrupt temperature changes that are characteristic of the glacial epoch. The large variety of proxy data available for the Holocene, including polar-ice and ocean-floor sediment cores, give some detail about the geographical extent, temporal variability and impact on the biosphere during the temperature fluctuations of the Holocene.

Despite the relative stability of the Holocene temperature, there are many indicators from within the palaeoclimate record that suggest hydrological variability occurred over subtropical regions. The relatively rapid change about 5,500 years ago that substantially reduced rainfall over Africa, and possibly over other subtropical regions, underscores the continuing potential for surprises. The strongest indicator is the abundant historical and cultural information referring to the impacts of drought and varying river flows that relate to the development of human settlements, especially over the Middle East.

There is evidence of multi-decadal temperature variations within the Medieval Warm Period and the Little Ice Age, at least over Europe and the North Atlantic. Significant regional shifts between cold and warm or wet and dry regimes have been identified from supporting proxy data collected from many locations around the globe and analysed. The proxy data do not always point to global coherence but the magnitudes of change have been sufficient to modify local ecosystems and have certainly modified human settlement patterns, sometimes beneficially and at other times adversely.

Many questions remain concerning the cause and extent of apparent decadal, centennial and millennial-scale fluctuations of the climate system. It has yet to be determined to what extent the variability is pervasive or of more sensitivity during glacial periods. There is also the question of whether many proxy markers (eg, ice raft debris) are only effective during glacial periods. Even over the recent 10,000 years, as the earth's axis of rotation has precessed through 180 degrees, we are unable to provide a physical explanation for the rapid pattern of warming between the last glacial maximum and the Holocene optimum, and the irregular cooling since.

Conditional instability

As we have noted, energy takes a circuitous route between when solar radiation is absorbed by the climate system to when it is later emitted to space as longwave radiation. Internal variability within the climate system occurs because there is not a balanced transfer as energy flows along the many pathways and, indeed, the pathways themselves are ever changing as the regions of the earth with excess and deficiency of net radiation change with season. As it moves within the climate system energy is exchanged between the underlying surface (both land and ocean) and the atmosphere, and these exchanges are regulated by surface temperature and wind speed. The changing temperature of the surface also represents changes in stored energy. Processes of the climate system that involve latent heat exchange (evaporation of water, formation of cloud droplets and snowflakes, and melting of snow and ice) are also conditional on prevailing temperature. Each event in which energy flow is constrained represents a build-up of energy and the development of conditional instability.

The most readily observed example of internal variability of the climate system is the weather systems of middle and high latitudes. These systems provide a more efficient mechanism than Rossby Waves for transporting energy towards the poles. Within the deep westerly winds that circle the earth over middle and high latitudes there is an ongoing process for developing baroclinicity (strengthening of the horizontal temperature gradients). As previously discussed, the Hadley Cells generate relative atmospheric angular momentum that has the effect of strengthening the westerly winds in the high troposphere. The need for thermodynamic balance causes the horizontal temperature gradient to strengthen. When baroclinicity reaches a critical level and there is a suitable trigger, instability occurs in the westerly flow and a middle latitude cyclone develops.

Prior to the formation of the middle latitude cyclone there is a generation of conditional instability. The increasing baroclinicity represents the juxtaposition of a very cold air mass and a warm air mass across the tight temperature gradient. As the baroclinicity increases so to does the available potential energy of the atmosphere because of the developing density differences of the air masses. The release of the available potential energy, or conditional instability, comes about as the cyclonic circulation becomes established; very cold air sweeps toward the equator and subsides and warmer air flows poleward and rises. Discrete weather systems, rather than continuous transport within Rossby Waves, are favoured for the efficient poleward transport of energy. The typical lifecycle of a middle latitude cyclone is therefore the characteristic timescale for middle latitude energy transport.

In the case of tropical storms, conditional instability develops as warm moist air accumulates in the tropical boundary layer. Deep convection is the process whereby the energy that is accumulating in the tropical boundary layer is distributed through the tropical troposphere. However, if the characteristics of the airflow do not favour deep convection then the accumulating energy is confined within the lower layers of the troposphere and conditional instability develops. Relatively strong vertical wind shear and/or low level temperature inversions that are typical of the Trade Wind regions are factors that will inhibit or prevent deep tropical convection, even when the sea surface temperature is relatively warm. Conditional instability builds up as heat and latent energy are transferred from the ocean to the atmospheric boundary layer but deep convection is inhibited.

Transient tropical weather disturbances, such as low-level waves in the easterly windfield, have local dynamically induced convergence in the airflow of the boundary layer. The migration of such a system across a region of accumulating boundary layer energy is often a sufficient trigger to establish deep convection and release conditional instability. Once established, the deep convection in the disturbance draws on the accumulated energy in the boundary layer of the local area. It is not uncommon for the disturbance to organise the local circulation into a tropical storm that is drawing in high energy boundary layer air from the surrounding region and ejecting air with high potential energy in the upper troposphere. Fully developed, tropical storms can organise the low-level wind field to draw in boundary layer air from over an area covering tens of thousands of square kilometres.

These two examples of developing conditional instability and its release are not unique in the climate system. They are familiar to us and they represent ongoing processes of the climate system that are of a lifecycle and regularity that can be readily observed. Each of the processes has its ranges of intensity and lifecycle duration that are regulated by the broader scale characteristics of the atmospheric circulation. For example, tropical storms are favoured in the summer hemisphere when sea surface temperatures are warmer and the supply of energy to the boundary layer is favoured; middle latitude storms are more frequent and intense when cooling of the polar troposphere is strongest and favours generation of available potential energy.

There are other processes within the climate system that favour transfer and accumulation of energy, and the development of other characteristic forms of conditional instability that have very long lifecycles. The five forms of conditional instability that will be reviewed below have timescales that range from the interannual to the millennial and are associated with characteristic processes within the climate system.

i. El Niño-Southern Oscillation (ENSO)

ENSO is cause for the most pronounced variability of the climate system after the seasonal cycle. Regional droughts and other areas with flood rain provide evidence of widespread and persisting changes to the atmospheric circulation during major El Niño events. It is now well established, based on the decade long (1985-1995) international Tropical Ocean Global Atmosphere (TOGA) research program of intense observations, that changes in the pattern of sea surface temperature across the tropical Pacific Ocean force global changes in the atmospheric circulation[132]. ENSO has been extensively studied over recent decades and yet the complex ocean-atmosphere feedback processes are still to be fully resolved. There has been an impetus to develop robust prediction methodologies in an attempt to provide early warning and mitigation strategies and to reduce the enormous social, economic and environmental impacts of El Niño events that are experienced worldwide.

The stress on the ocean surface by the normal easterly Trade Winds blowing across the tropical Pacific Ocean establishes the characteristics of conditional instability. The elevated sea level and depressed thermocline over the western Equatorial Pacific Ocean are maintained only while the Trade Winds persist. The elevated sea level and depressed thermocline are characteristic conditions of the ocean surface layer that are necessary to be present prior to El Niño development. Anomalous surface equatorial westerly winds that dramatically reduce the wind stress on ocean surface over the western Pacific Ocean have been identified as important for the initiation and

---

132 See the earlier section describing ENSO.

subsequent development of major El Niño events. The westerly winds apparently occur without any external stimulus to the climate system.

Once an El Niño is established, and warmer than normal waters have spread across the equatorial central and eastern Pacific Ocean, there is an enhancement of the surface heat and moisture exchange between the ocean and atmosphere, and an enhancement of the meridional transport of energy. Correlations and time filtering of the observed atmospheric angular momentum and sea surface temperature fields show coherent, global-scale propagation of angular momentum anomalies from the tropics to high latitudes[133]. Studies, also based on observations, show that there is an enhancement of poleward transport of energy by the atmospheric circulation during these events[134].

The period of variability of ENSO is characteristically on the interannual timescale. The studies further demonstrate that variations in tropical sea surface temperature associated with ENSO force a response in the location of the Rossby Waves, the strength of the westerly winds and the locations of middle latitude cyclone development and their subsequent trajectories. Through the shift in favoured location of tropical convection, the subsequent interactions with the Hadley Cell circulations and strengthening of the westerly airflow, ENSO also contributes to variability of middle and high latitude climate with a characteristic interannual timescale.

There is some evidence to support the view that El Niño events are more frequent and are more intense when the tropical oceans are warmer than normal. There has been a preponderance of El Niño events since the mid-1970s when tropical oceans most recently have warmed. The alternative view, that tropical oceans are warmer because of the more frequent El Niño events, still relies on a source of energy to heat the surface waters of the equatorial Pacific Ocean.

ii. The ocean gyres and upwelling
The ocean gyres of the subtropical oceans circulate relatively slowly. The circulations are largely driven from the wind stresses of the tropical Trade Winds and the middle latitude westerly winds. There is also the suggestion that the ocean gyres derive some of their circulation energy from the pumping action associated with middle latitude cyclones. In addition to driving the circulation, the wind stress causes an Ekman Drift towards the centre of the gyres and sea level is higher in the central regions of the subtropical gyres than on the periphery.

133 Dickey, J.O., S.L. Marcus and O. de Viron, 2003. Coherent interannual and decadal variations in the atmosphere-ocean system. Geophys. Res. Lett., vol 30, No. 11 (27).

134 Trenberth, K.E., D.P. Stepaniak and J.M. Caron, 2002. Ibid.

Variations in the circulation of the ocean gyres are part of the complex interactions between the atmosphere and the oceans. Meteorological records identify multi-decadal variations of wind strength and in pressure patterns, particularly for the northern hemisphere[135]. Although the local linkage between wind stress and ocean currents is relatively well understood, the impacts of multi-year variations of seasonal wind are less understood, particularly taking into account the mass inertia of each of the large ocean gyres. Feedbacks from the ocean circulation to the atmosphere, including the persisting impacts of temperature anomalies over the middle latitudes, are yet to be resolved. There is emerging evidence that decadal scale dry periods in some regions, such as the 1930s 'Dust Bowl' drought event of North America, are linked to persisting anomalies of regional sea surface temperature.

Local sea surface temperatures are moderated by the oceanic transport of heat and local upwelling but it is the local sea surface temperature that regulates the rate of heat and latent energy exchange between the oceans to the atmosphere. The comprehensive satellite observations of sea surface temperature over the past three decades, together with ship-based and drifting buoy measurements, have identified temperature anomalies that extend through the surface mixed layer and persist for a number of years. Temperature anomalies of the North Pacific Ocean have been tracked for a number of years as they move with the flow.

Some information about local ocean temperature changes associated with subtropical gyres and upwelling has been obtained through interpretation of ocean-floor sediment cores. A core recovered from the Santa Barbara Basin of the northeast Pacific Ocean identified temperature variability on multi-decadal and centennial timescales over the period from 11 to 3 thousand years ago[136]. The time resolution of the core was only 8 years and nothing could be gauged about the shorter interannual to decadal timescales. The plankton species analysed represented both the surface water and deeper water near the thermocline. Analysis of the changing ratio of oxygen isotope (^{18}O) has led researchers to believe that cooler waters at the end of the last glacial period gave way to warmer waters that were at their peak warmth between 7 and 4 thousand years ago. The magnitude of

135 The North Atlantic Oscillation (NAO) is an interannual to decadal fluctuation in the tracks of storms from the Atlantic to Europe that is linked to changes in the westerly wind strength and regional pressure gradients. There is a similar North Pacific Oscillation (NPO) and both are suggested to be linked to an Arctic Oscillation (AO).

136 Friddell, J.E., R.C. Thunell, T.P. Guilderson and M. Kashgarian, 2003. Increased northeast climate variability during the middle Holocene. Geoph. Res. Lett., vol 30, No. 11 (14).

variability was much greater over the interval of maximum warmth, which also corresponded to maximum vertical temperature stability of the surface water.

In modern times it has been found that the surface water temperature of the Santa Barbara Basin responds to the strength of local upwelling. However the seasonal and interannual characteristics of upwelling are not resolved in data from the ocean-floor sediment core. There are two plausible explanations for the multi-decadal to centennial fluctuations in ocean surface temperature over the region and both processes may be contributing together. Firstly, variations of the atmospheric circulation and local winds on these longer timescales will impart a similar response time to the strength of ocean upwelling. Secondly, variations in the atmospheric circulation will impact on the wind-driven surface currents of the oceanic gyres, including the strength of the cold southward flowing current offshore from the West Coast of North America, on a similar timescale.

The potential for in-phase or out-of-phase linkages between the strength of the atmospheric circulation and the subtropical ocean gyres is real. The palaeoclimate indicators suggest that ocean surface temperatures in the regions of the ocean gyres do vary on multi-decadal and longer timescales. Reliable wind records are not sufficiently long, nor do they have a good spatial coverage but there are a number of circulation proxies based on surface pressure indices that also indicate multi-decadal fluctuations. The origins of the multi-decadal circulation fluctuations have not been identified.

iii. Bottom Water formation

The thermohaline circulation is a millennial scale overturning of the ocean brought about by cooling of surface waters of relatively high salinity to below 0°C over polar regions during winter. At these sub-zero temperatures the ocean surface waters are very dense, particularly when compared with the warmer and more buoyant tropical and subtropical water. The density increase during cooling is further assisted by the ejection of salt into the surface water where sea ice is forming and thickening. There are unresolved issues about the relative roles of cooling and salt enhancement versus the need for wind forcing to overcome the natural stratification of the oceans. Notwithstanding these, the polar regions have substantial net radiation deficiency during winter. The regular surface water cooling and sea ice formation provide the appropriate conditions for bottom water formation, if not the actual forcing.

Evidence of historical changes in the rate of bottom water formation comes from ocean-floor sediment cores. An ocean core recovered from the subpolar North Atlantic Ocean has been analysed to provide a record of carbon isotopes over the period of the Holocene[137]. The proportions of ^{13}C at various times in the record have been linked to the rate of North Atlantic Bottom Water formation. The isotope record suggests that there has been a slow reduction in the rate of formation over the past 6,500 years. The record also identifies periods of marked reduction in formation around 9.3, 8.0, 5.0 and 2.5 thousand years ago. Other reductions, of smaller magnitudes, occurred more frequently. A high relative abundance of haematite stained grains in a nearby ocean-floor sediment core indicates significant intrusion of cold, fresh, ice bearing surface water from north of Iceland into the North Atlantic during periods of reduced bottom water formation.

Ocean-floor sediment cores indicate that the major fluctuations of bottom water formation are consistent with ocean variations over the entire Atlantic Basin. Some palaeoclimate data point to near synchronous climate changes over Europe and North Africa but apparently to a lesser extent elsewhere.

There is uncertainty about the causes for variability in the rate of North Atlantic Bottom Water formation. The major reversion to very cold conditions, during the Younger Dryas period that lasted about 1,000 years, has been attributed to a flood of fresh water through the St Lawrence drainage and into the North Atlantic. Explanations for slow freshening of the surface layers of the North Atlantic Ocean over more recent times have been linked to excess of precipitation over evaporation but such linkages are yet to be substantiated through rigorous evaluation. Freshening of the surface layers would weaken the overturning because of the reduction of meridional density gradient and the introduction of further vertical density stratification. A feedback involving the strengthening of the wind-driven Gulf Stream, and enhancement of the import of saline subtropical surface waters, has been advanced, but is also a process that needs further investigation.

Notwithstanding a lack of knowledge about its source, there is very strong evidence that the rate of North Atlantic Bottom Water formation does fluctuate over centennial to millennial timescales and that there are impacts on regional climate. However it is not entirely clear whether the regional climate effects are an outcome of the atmospheric changes that are contributing to the changing bottom water formation, or they are a direct outcome of the changing bottom water formation itself.

---

137 Oppo, D.W., J.F. McManus and J.L. Cullen, 2003. Deepwater variability in the Holocene epoch. Nature, vol 422, pp277–278.

iv. Antarctic upwelling

The surface waters surrounding Antarctica have their origins in the middle depths of the Southern Ocean[138]. The upwelling is a direct result of the middle latitude westerly winds of the surface and their frictional stress on the surface waters of the Southern Ocean. The equatorward directed Ekman drift of the surface water causes upwelling on the poleward margins of the circumpolar current and very cold water from depth is brought to the surface. The very cold surface waters of the circumpolar current isolates the waters around Antarctica from the warmer surface waters of the South Pacific, Indian and South Atlantic Oceans. Evaporation from the cold Antarctic waters is low and there is a tendency for the surface water to be freshened by precipitation. The tendency for freshening is countered by wintertime salt injection during the formation of sea ice on the continental margins.

The surface Ekman drift and upwelling of cold water respond directly to the speed of the prevailing westerly winds, albeit with a delay due to the inertia of the water mass. There is more upwelling of very cold sub-surface water when the winds are stronger. The stronger upwelling works toward maintaining both the very cold temperatures and the salinity of the surface waters around Antarctica and both of these factors are conducive to the more rapid formation of Antarctic Bottom Water. There is a direct linkage between the strength of the high latitude westerly winds and formation of Antarctic Bottom Water.

At the Antarctic Convergence, to the north of the Antarctic Circumpolar Current, the cold Antarctic surface waters generated by the Ekman Drift plunge under the relatively warm subtropical waters. The magnitude of the upwelling and ocean overturning is also directly related to the speed of the surface westerly winds. In turn, the speed of the westerly winds is an outcome of the rate of generation of relative atmospheric angular momentum through atmosphere overturning associated with the Hadley Cells. Research is needed to identify how changing upwelling and overturning, and impacts on energy processes, feed back to the climate system or is damped with time.

v. Ice sheets and mountain glaciers

A dominant climate fluctuation over the last millennium has been the worldwide advance of mountain glaciers commencing about the 14th century (although probably earlier in sub-polar regions) and their almost synchronous retreat since the middle of the 19th century. Records of coastal sea-ice occurrence around Iceland suggest that the glacial advance and retreat was synchronous with changes in the polar

---

138 See the Section: Southern Ocean upwelling and Figure 21.

ice sheets, at least for the northern hemisphere. This near millennial-scale fluctuation is the most recent manifestation of a series of fluctuations for which there is some evidence through the Holocene, although global data for estimating the timing and magnitude of earlier events is limited and inconclusive. There have been other regional climate fluctuations of shorter duration but the glacial advance and retreat describe a broad envelope of change synchronously effecting both hemispheres.

The accumulation of polar and mountain ice mass during glacial advance represents a persisting global energy imbalance. Here we recall that the latent energy released to the atmosphere during snow formation is in excess of the latent energy extracted from the oceans during evaporation of water vapour. The accumulation of snow represents a net loss of energy from the climate system to space. Similarly, the melting of polar and mountain ice represents an uptake of energy by the climate system as latent energy is stored in the melt water that flows back to the oceans.

The conventional explanation for the accumulation of snow and the advance of ice sheets relates to a reduction of solar insolation and snowmelt over the polar regions during summer. We can link the changing insolation over polar regions to the earth orbital characteristics (precession, tilt and eccentricity). However, this is not a satisfactory explanation when fluctuations of shorter timescale than the 22,000-year precession period are considered.

As has been emphasised, the polar regions are regions of annual net radiation deficit and rely on the importation of energy by the atmospheric and ocean circulations to maintain their seasonal temperature regime, especially during winter. A reduction of the poleward transport of energy from the tropics is a very effective mechanism to enhance the energy deficit over middle and high latitudes. The processes that regulate meridional energy transport cannot be taken as fixed and are clearly important for modulating the overall energy balance over high latitudes.

Over polar regions the emission of terrestrial radiation is also regulated by the prevailing surface temperature and the temperature of the troposphere. A change in the rate of importation of energy from the tropics by the atmospheric circulation will not immediately change the rate of terrestrial emissions to space. It will affect the surface temperature and the rate of snow accumulation or snow melting, depending on whether the rate of importation of energy is reduced or enhanced. An increase in the rate of energy transport from the tropics will tend to warm the underlying surface and increase the rate of snow and ice melt. A decrease in the rate of import of energy from the tropics will result in cooling of the underlying surface and favour the accumulation of snow. Whether snow melts or accumulates during a respective increase or decrease of poleward energy transport depends on whether the surface temperature is below or above 0°C. Accumulation of snow and build-up of polar ice sheets is favoured when there are large areas where the surface temperature is below 0°C.

Sensitivity to solar forcing

In the earlier discussion about solar influences on climate it has been noted that the IPCC considers changes in solar irradiance to be a 'radiation forcing' whose impact is equally distributed across the surface of the globe. The IPCC treatment underestimates the sensitivity of the climate system to variations in solar irradiance. The specification of globally uniform solar insolation will allocate too much to the polar regions and not enough to the tropics. At the top of the atmosphere the annual solar insolation over polar regions is less than half that over the tropics. In addition, 80 to 90 percent of solar insolation incident on snow and ice is reflected to space. Overall, the IPCC approach underestimates the impact of variations in solar radiation on the energy available to the climate system because of the underestimation of insolation absorbed in the tropics and the overestimation of that reflected to space.

In reality, a change in solar irradiance is focused over the tropics and alters the energy accumulating in the tropical oceans. The climate system responds to the rate at which energy is accumulating in the surface layers of the tropical oceans and the tropical sea surface temperature. These regulate the exchange with the atmospheric boundary layer, the rate of convective overturning of the tropical Hadley Cells, and the rate of meridional transport of energy by the Rossby Waves and weather systems. However there is no immediate impact on the emission of terrestrial radiation to space. The surface temperature and the temperature lapse rate in the troposphere regulate the latter. The large thermal capacity of the ocean surface layer ensures that neither the ocean surface nor the troposphere temperatures are immediately affected by variations in solar radiation. As occurs during an El Niño event, the increased ocean energy exchanged across the sea-air interface to the atmospheric boundary layer will mainly be as latent energy.

There is an ongoing excess radiation over the tropics and deficits over the polar regions, and any change to the solar irradiance will, therefore, be reflected in the rate of poleward energy transport. A change in the meridional transport of energy to the middle latitudes and polar regions by the atmospheric circulation will have a direct impact on climate characteristics of the earth's surface. The surface temperature over this part of the hemisphere is sensitive to the rate of atmospheric transport because it is a region of net radiation deficit. An increase in the rate of energy transport will raise surface temperature and a decrease will lower surface temperature. There will also be a capacity to melt or accumulate snow and ice depending on whether the energy transport is increased or decreased. Melting or accumulation of snow and ice are critically dependent on the location of the 0°C isotherm.

The direct measurements of solar irradiance are inconclusive. However, the response by the climate system accords with a recent increase in solar irradiance over the past two decades. The surface temperatures of the warmest waters of the tropical oceans have been constrained to about 30°C but the average surface temperature has increased. As we would expect, because of the need to maintain buoyancy in the deep convective clouds, the temperature of the tropical troposphere has not increased[139].

The increase in average sea surface temperatures has meant that surface emission of terrestrial radiation has also increased. Most of the terrestrial radiation emitted from the surface is absorbed in the troposphere by greenhouse gases, clouds and aerosols but part does get transmitted directly to space through the atmospheric window, the wavelength bands where there is no absorption. Observations by satellite of longwave radiation to space confirm that across the tropical latitudes 20°N to 20°S there has been an increase in longwave emission over the past two decades[140]. Although the temperature of the troposphere has not increased and there has been a small increase in the radiation to space through the atmospheric window, the increase in longwave radiation to space has been ascribed to a decrease in tropical cloudiness. Independent analysis of satellite derived longwave emission data support the view of overall decreased tropical cloudiness and point to a strengthening of the Hadley Cell circulations, including more intense regions of equatorial convection and drier and less cloudy equatorial and subtropical subsidence regions[141].

* * * * * * * *

The IPCC proposition of a stable climate system regularly changing through the annual cycle for about 900 years prior to the onset of industrialisation does not accord with cultural and historical records

139 Spencer, R.W. and J.R. Christy, 1990. Ibid.
140 Wielicki, B.A., T. Wong, R.P. Allen, A. Slingo, J.T. Kiehl, B.J. Soden, C.T. Gordon, A.J. Miller, S-K. Yang, D.A. Randall, F. Robertson, J. Susskind and H. Jacobwitz, 2002. Evidence of large decadal variability in the tropical mean radiative energy budget. Science, vol 295, pp841–844.
141 Chen, C., B.E. Carlson and A.D. Del Genio, 2002. Ibid.

nor with a significant body of palaeoclimate evidence. There is a very strong signal of internal climate variability on the interannual timescales (ENSO) and climate records point to prolonged anomalies that are believed to have their origins in fluctuations in ocean temperature patterns. Ocean circulations, both the subtropical gyres and the meridional overturning, have large mass inertia and long timescales associated with their circulations. Although the ocean circulations themselves transport only a relatively small proportion of the total energy carried from the tropics, any variation in the circulation will be reflected in changed surface temperature patterns that regulate the intensity and patterns of atmospheric transport.

The evidence of slow warming of the earth's surface over middle and high latitudes and melting of ice mass since the middle 19th century is compelling. The melting of mountain glaciers and polar ice mass supports the view that there has been a global net radiation imbalance at the top of the atmosphere that persisted through the period. The cause of the radiation imbalance is alternatively a small increase in solar irradiance or a decrease in terrestrial emissions. The satellite observations and the pattern of changing atmospheric circulation, including strengthening of the Hadley Cell circulations, over the past to decades are consistent with increased solar irradiance. Such a conclusion conforms qualitatively to observations of the solar characteristics and to isotope data that reflect a recent increase of solar irradiance. However, a reduction of the thermohaline circulation and a commensurate increase in the poleward transport of energy by the atmospheric circulation cannot be discounted.

Computer Modelling of the Climate System

The IPCC projections of future global warming are dependent on the efficacy of computer models and their ability to realistically represent the processes of the climate system. IPCC claims that the ability of computer models to provide 'useful projections of future climate' has improved because of the demonstrated performance on a range of space and time scales. If such claims have foundation then the computer models must be sensitive to internal variability on timescales that are observed and responsive to 'external' forcing consistent with natural forcing of the climate system. The IPCC draws attention to positive developments in computer modelling, including:

- over long simulations the models are stable, they adequately represent a variety of statistical characteristics of the climate, and they exhibit only limited internal variability;

- simulations for the 20th century with imposed forcing that mimics natural forcing (solar irradiance variations and vulcanism) and anthropogenic forcing (aerosols and increasing greenhouse gas concentrations) project a pattern of global mean surface air temperature that has many of the characteristics of the observed temperature pattern for the period.

It is true that computer models of the climate system are undergoing a rapid rate of development. The increasing speed of computers is being translated directly to improved resolution of the atmospheric and ocean circulations, and to improved representation of local processes, including those associated with clouds, sea ice and exchanges between the underlying surface and the atmosphere. Also, many uncertainties relating to atmosphere and ocean processes are being resolved through research and these advances are being incorporated into computer models.

Notwithstanding the improvements over recent decades, computer models of the climate system are still very rudimentary. There are fundamental limitations to our knowledge about the ocean circulations and their variability, the so-called flywheel of the climate system, whose interactions with the atmosphere are important for determining internal variability of the climate system. The representation of ocean circulations and energy transport in the computer models is of particular importance in assessing the relevance of projections of future climate.

The scope for internal variability of the climate system has been discussed in the previous section. The observed variability of ocean circulations on interannual to decadal timescales is recognised as the source of climate variability with concomitant disastrous impacts on communities through drought and flood. There is also evidence that the climate has varied on centennial to millennial timescales, with advancing and receding cold over middle and high latitudes and unexplained drying of many subtropical land areas over the past 5,000 years. It is surprising, therefore, to find that IPCC claims that lack of internal variability in computer models is a positive attribution. In the past, because of the need to avoid computational instability, it was necessary to construct computer models in such a way as to avoid significant internal variability. Our knowledge of ocean and atmosphere circulations and their interactions suggests that lack of internal variability of computer models is not a positive attribute but a limitation of the formulations of processes and damping of potential non-linear interactions.

IPCC does acknowledge that the computer models do continue to have a number of limitations that affect their performance. Assumptions about 'feedback processes', especially involving clouds and their interactions with radiation and water vapour, are recognised to impact on the sensitivity of the climate system to external forcing. Despite recent advances, some of the energy exchange processes, including the formation of sea-ice, are still poorly represented in computer models. Also, there is an inability to account for the different temperature trends observed at the surface and in the middle troposphere since 1979, notably the troposphere does not exhibit warming while the surface temperature is increasing.

It is important to objectively assess the current status of computer models. We need to be confident that the computer models are sufficiently developed to adequately represent the climate system and realistically project future climate states. The apparent ability of computer models, with seemingly plausible natural and anthropogenic forcing, to represent the global surface temperature pattern of the 20th century is cited by IPCC as one of its benchmarks. On the basis of coherence between statistics of the computer models and the observed climate, and the ability of models to replicate the 20th century climate, IPCC has expressed confidence in the ability of models to project future climate. It is against the IPCC benchmarks that the computer models and their ability to project climate into the future will be assessed here.

Development of Computer Modelling

Computer models of the climate system have their origins in weather forecasting. In 1922, the English meteorologist L.F. Richardson proposed a method for weather prediction by computations. Richardson's method was based on a well-known set of equations of motion for fluid flow as applied to the atmosphere on a rotating earth. No more than a demonstration of the potential of the methodology could be given because of the large number of computations that were required. It is because of the number of calculations involved that the early development of numerical weather prediction was slow and only advanced in parallel with advances in computing technology.

The development of weather prediction was one of the first applications for the large-scale 'computing machines' developed in the late 1940s and early 1950s. The first 'models' were very simple affairs to predict changes in surface pressure patterns over relatively limited regions of the globe. They had no representation of small-scale physical processes and the subsequent development of pressure patterns was reliant on the energy distribution represented by the initial state. However, as computers became more powerful there was more opportunity to better represent the atmosphere and include more of the physical processes that contribute to its ever-changing flow.

By the early 1970s weather prediction models that covered the full global domain had been developed. These models could resolve weather systems a few hundred kilometres across; they discriminated more than a dozen separate levels in the vertical, and had representations for a range of physical processes, such as clouds (for predicting precipitation) and boundary layer friction. The early numerical weather prediction models lacked energy sources and sinks and, as the available potential and kinetic energy represented in the initial state of the system dissipated, the weather systems lost intensity. Also, the inability to adequately specify initial conditions, and the generation of mathematical 'truncation errors' and their propagation, seriously compromised the predictions within a day or so.

It was during the early 1970s that weather forecasting models began to be adapted for climate purposes. Here, the objective was not to replicate the evolution of individual weather systems but to replicate seasonal and annual statistics (the 'climate'). The main change to the models, if the simulations were to cover years and longer, was to include balanced energy sources and sinks. Initially the representation of the energy source was rudimentary – climatologically specified land and ocean surface temperatures and eddy transfer of heat, moisture and momentum based on the surface conditions and the wind, temperature and humidity of the air in the layer next to the surface. The energy sink was represented by radiation cooling of the atmosphere. Radiation transfer theory for estimating radiation fluxes in the atmosphere was well known but the limitations of computing power prevented very elaborate representation, especially when clouds were also taken into account.

By the early 1980s the ocean was being represented by stationary layers (the so-called 'slab ocean') into which solar insolation was partitioned and for which adjustments were made to represent estimated heat transport by the ocean currents. It was this class of model that was the basis for the early 'Double CO2' experiments. In these experiments, a computer model was integrated for a very long time to reach a stable 'climate' state (the control). The representation of radiation was then altered to simulate the doubled concentration of atmospheric carbon dioxide and again integrated from the initial conditions to a stable 'climate'. The difference in the statistics of the 'climate states' between the two simulations represented the impact of doubling atmospheric carbon dioxide concentration and became the basis for early predictions of global warming, for example as used at the Villach Conference in 1985.

Over the past two decades, computer models representing the climate system have become more elaborate as the development of computing power has accelerated. Important advances include:

- a better dynamic representation of the ocean as a fluid in motion,

- representation of the polar ice sheets, including sea ice and its seasonal advance and retreat,

- representation of the land surface with a capacity to store precipitation in the soil, simulate run-off, regulate evapotranspiration, and change local albedo with changing vegetation or snow cover,

- representations of clouds and their characteristics (altitude, depth, etc) that change with local atmospheric conditions, and

- representation of atmospheric radiation that changes with cloudiness and cloud type.

The increased complexity of important climate processes incorporated in the computer models provides the opportunities for a more accurate representation of the climate system. Indeed, with contemporary solar irradiance and greenhouse gas concentrations, computer models are better able to characterise the large-scale features of earth's climate. However, modellers still caution that many of the small-scale ocean-atmosphere exchanges, etc., are only crudely approximated. Many of the small-scale processes respond non-linearly to the changing atmospheric and ocean conditions and it is not known how errors propagate through the model to affect the outcome.

Extended simulations of the atmospheric circulation, forced by prescribed patterns of sea surface temperature, have given early credence to the belief that computer models had the potential to simulate the complex climate system. In these simulations, 'climate' is represented by the seasonal and annual means of atmospheric characteristics. In many respects, 'climate' is still validated against statistics of atmospheric characteristics, mainly because it is from the atmosphere that most data suitable to construct benchmarks of performance are available.

The integrity of the atmospheric component of climate simulations, that is, atmospheric general circulation models, has been established by comparing the performance of different models under the same forcing conditions. Within the framework of the Atmosphere Model Intercomparison Project (AMIP), various models have been forced by the same set of sea surface temperature patterns covering the decade 1979–1988 and their respective outputs compared to atmospheric observations over the period. The results of the model intercomparisons have been reported in the literature and identify characteristic weaknesses of individual models and systematic weaknesses across models.

One measure of model performance is the time history of atmospheric angular momentum. We have previously noted that the overturning Hadley Cells of the tropics generate relative atmospheric angular momentum. The ability of the computer model to represent atmospheric angular momentum is one measure of the model's ability to distribute energy from the boundary layer through the troposphere. In a comparison between 23 atmospheric models, the decadal mean atmospheric angular momentum was well simulated and 10 of the models were within 5 percent of the observed value. Half of the models represented the median amplitude of the seasonal cycle within 15 percent of the observed value for the period. However, the models were generally deficient in their ability to represent interannual variations of atmospheric angular momentum, especially underestimating the 1982–83 El Niño event and overestimating the 1986–1987 El Niño event[142]. These comparisons suggest that the atmospheric component of the computer models do not respond well to changing forcing from the ocean.

Experiments have been carried out with computer models to identify the sensitivity of the climate system to specified sea surface forcing. In particular, 'El Niño experiments' were carried out to determine if there is a systematic response of rainfall, temperature and winds over other parts of the globe when warm anomalies of sea surface temperature are prescribed over the equatorial Pacific Ocean. In response to such forcing, the computer models consistently reproduce, at least qualitatively, many of the observed characteristics of the El Niño response.

Despite their apparent simplicity, earlier computer models with slab representations for ocean energetics were able to reproduce many of the large-scale seasonal and annual statistics of the atmosphere (the 'climate'). This was an encouraging start but the results were heavily constrained because of the treatment of solar radiation and assumptions about partitioning of energy within the slab ocean. Better utility was achieved by including a dynamic representation of the ocean, where sea surface temperature responded more directly to seasonal solar insolation as modified by clouds and ocean transports.

The inclusion of a dynamic representation of the ocean circulation (similar to that of the atmosphere) led to less constrained interactions between the components of the climate system. Computer models representing the coupled ocean-atmosphere system initially had a tendency, over decades and centuries, for 'climate drift' from the initial state as they accumulated energy and warmed. The early coupled models initially relied on 'flux adjustment' (prescribed adjustments to the ocean-atmosphere exchange) to limit climate drift. In more recent formulations the computer models do not rely on flux adjustment and there is only limited climate drift.

The uncertainty and sensitivity of prescribed non-linear processes, and feedback between components of the system, have the potential to confound computer model predictions. For example, an error of only 1.3 percent in the specification of the global mean albedo, or its change as a result of non-linear interactions between processes, will alter solar insolation absorbed within the climate system by about 4 W/m^2. This represents an inadvertent forcing of the climate system that is equivalent to the 'radiative forcing' from doubling the

142 Hide, R., J.O. Dickey, S.L. Marcus, R.D. Rosen and D.A. Salstein, 1997. Atmospheric angular momentum fluctuations during 1979–1988 simulated by global circulation models. J. Geophys. Res., vol 102, No. D14 pp16, 423–16, 438.

concentration of atmospheric carbon dioxide. Changes to sea-ice extent, land surface characteristics and cloudiness each affect global albedo. Unwitting feedbacks that change internal processes, especially if the feedback is sensitive to temperature, will mask or erroneously amplify the direct radiative forcing attributed to increased greenhouse gas concentrations.

Local water vapour concentration in the atmosphere is a function of temperature, cloud formation and atmospheric circulation. Of special interest, the maximum amount of water vapour that the atmosphere can hold increases rapidly with temperature, and water vapour is a primary greenhouse gas. If the atmosphere warms with time as a result of increasing concentrations of anthropogenic greenhouse gases, or for other reasons, then the resulting additional water vapour will further enhance the greenhouse gas forcing. This amplification of the initial warming is referred to as a positive feedback.

In computer model simulations that include a doubling of atmospheric greenhouse gas concentration, water vapour is consistently cited as the most important positive feedback process and amplifies the climate response[143]. Water vapour feedback alone approximately doubles the warming from what it would be for fixed water vapour. Furthermore, water vapour feedback acts to amplify other feedbacks in models, such as cloud feedback and ice albedo feedback. However, in assessing the veracity of the positive water vapour feedback proposal it is essential to be mindful of the observations of tropospheric temperature during the 1980s and 1990s made independently using satellites and balloons. Although this was a period of systematic surface temperature increase, attributed by IPCC to greenhouse gas forcing, there was no appreciable temperature change in the troposphere. That is, water vapour feedback was definitely not contributing to nor amplifying the surface temperature change over this period when observations have been available.

One of the contentious issues in computer modelling of the climate system is how to represent cloud-radiation processes and the effectiveness of cloud-radiation feedback. In experiments to predict the change in climate from a doubling of atmospheric carbon dioxide concentration, the top of the atmosphere change in net radiation (radiative forcing) due to cloud-radiation feedback processes for most models is less than 3 W/m^2, well within the range of observational uncertainty. However, there is a large discrepancy between models as to sign, magnitude and share between longwave and shortwave components of the feedback radiative forcing. This disagreement (amplified by water vapour feedback) reflects the sensitivity of response to computer model formulation and is a major cause for the

143 IPCC, 2001. Ibid, p425.

wide range of the global surface temperature response (ie, sensitivity) by the different computer models. As the IPCC explains, "there has been no apparent narrowing of the uncertainty range associated with cloud feedbacks in current climate change simulations"[144].

There are still many uncertainties in the specification of processes represented in computer models of the climate system. For some processes, critical values cannot be accurately measured. As a consequence, they have either been estimated or are formulated from the results of separate model studies. Studies show that the formulation of sensitive processes is crucial to the climate response of the computer model. Moreover, unless there is a sound, physically based preference for a particular formulation then the choice, and the climate response, are arbitrary[145].

An example of how new data can radically alter seemingly plausible assumptions comes from satellite measurements that have identified a semi-annual oscillation of reflected shortwave radiation over the tropics[146]. The amplitude of the oscillation is near 6 W/m^2 and has been linked to seasonally changing tropical cloudiness. These new data both provide an additional benchmark for computer model performance and also assist in better understanding and specifying the processes within the climate system. Prior to the analysis of the satellite data there was no indication that the climate system was subject to a natural semi-annual radiative forcing of a magnitude that exceeds the anthropogenic forcing expected after many decades. The semiannual forcing was not previously recognised and is neither reflected in the formulation of computer models as a natural forcing nor is it reflected as a component of the internal variability of computer models.

Performance of computer models

Several of the computer models examined by the IPCC reproduce variability of global surface air temperature on the interannual timescale that is of similar magnitude to that of the climate system[147]. However, the source of variability is only weakly linked to an ENSO-like phenomenon in the climate simulations. Variability of sea surface temperature over the equatorial Pacific in the computer models is weaker than observed[148], although the variations are sufficient to be reflected in altered tropical precipitation patterns and the strength of

144 IPCC, 2001. Ibid, p431.
145 Williams, K.D., C.A. Senior and J.F.B. Mitchell, 2001. Transient climate change in the Hadley Centre models: the role of physical processes. J. of Climate, vol 14, pp2659–2673.
146 Wielicki, B.A., et al, 2002. Ibid.
147 IPCC, 2001. Ibid, Fig. 12.2, p704.
148 IPCC, 2001. Ibid, p504.

monsoons. The accuracy of computer models that have been constructed specifically to simulate and predict ENSO is not yet reproduced in the existing lower resolution models for simulating climate change.

The Pacific Decadal Oscillation has been identified as the main mode of variability on decadal timescales and has its focus in the North Pacific Ocean, away from the tropics. Several computer models are able to reproduce aspects of this pattern of variability, although with limited fidelity. Despite the limitations in reproducing the Pacific Decadal Oscillation, the variance over decadal timescales of surface temperature simulated by the computer models is also similar to that observed in the climate system.

Beyond the decadal timescale there is no unambiguous benchmark against which the variations of the computer models can be judged. Computer models do not reproduce centennial to millennial-scale variability that has been identified in many palaeoclimate records and supported by the historical and cultural records from the North Atlantic-European sector of the Northern Hemisphere. The very long simulations made with several computer models would suggest that the climate system has limited variance beyond the multi-decadal timescales. The lack of variability in the computer models is surprising, given the complexity of the processes and their interactions within the climate system. The lack of internal variability suggests that the computer models are not sensitive to the internal processes that are qualitatively observed in the climate system. For example, the unforced models show no evidence of variations in the production of North Atlantic and Antarctic Bottom Water, nor the multi-decadal to centennial scale variations in surface water temperatures identified in ocean-floor sediment cores.

The physics of longwave radiation transfer in a cloudless atmospheric column dictate that an increased concentration of greenhouse gases will reduce the net longwave radiation emitted at the earth's surface. That is, energy will be retained in the underlying land or ocean surface. Over tropical oceans the retention of energy would be expected to cause a slow rise in ocean surface layer heat content and, ultimately, sea surface temperature. Over tropical oceans, where there is excess solar radiation and the prevailing temperatures are already warm, evaporation is a very effective process to transfer heat to the atmospheric boundary layer without need for a significant increase in temperature.

The Coupled Model Intercomparison Project (CMIP) provided an opportunity to compare the performance of different computer models and to compare their respective performances under similar conditions of greenhouse gas forcing. The international project was coordinated within the framework of the World Climate Research Programme and was managed within the Program for Climate Model Diagnosis and Intercomparison at the Lawrence Livermore National Laboratory in California, USA.

CMIP had two phases. The first phase was collection and comparison of statistics from 'control runs' of the respective computer models. During these 'control run' simulations the greenhouse gas concentrations and other forcing were held constant. The second phase was the collection and comparison of statistics from computer simulations where atmospheric carbon dioxide was increased by a specified one percent per year. The rate of increase was not claimed to be representative of future emission rates but was a realistic specification for comparison between models.

i. CMIP (Phase 1)[149]

The control simulations of fifteen computer models from 12 institutions have been compared. Some of these models relied on flux adjustment (prescribed values for heat and moisture exchange between the oceans and atmosphere) to limit climate drift but models without flux adjustment generally had only small climate drift. Additional information about the control run intercomparisons has been obtained from the evaluation of models during the second phase[150].

Many characteristics of the climate system are well specified, both qualitatively and quantitatively, in the control simulations. Near surface air temperature is particularly well modelled over the tropics and middle latitudes but disparities with observations are evident over the ice sheets of Antarctica and Greenland. The geographical distribution of precipitation is realistic within the models but there are systematic biases. In particular, there is an underestimation of peak rainfall over the tropical oceans and many models fail to capture the rainfall of the South Pacific Convergence Zone, instead erroneously specifying a band of precipitation across the Pacific Ocean south of the equator.

The tropospheric temperatures of the models tend to be cooler than observations suggest, especially in the high troposphere of middle and high latitudes. In the tropics the middle tropospheric temperatures are cooler than observed while the upper troposphere and stratosphere are warmer than observed. The specific humidity of the tropical

---

149 Lambert, S.J. and G.J. Boer, 2001. CMIP1 evaluation and intercomparison of coupled climate models. Climate Dynamics, vol 17, pp83–106.
150 Covey, C., K.M. AchutaRao, S.J. Lambert and K.E. Taylor, 2000. Intercomparison of present and future climates simulated by coupled ocean-atmosphere GCMs. PCDMI Report No. 66. Lawrence Livermore National Laboratory, USA.

troposphere is too low, while over middle latitudes the moisture is too high, especially in the low troposphere.

The models capture the position and intensity of the jet streams of each hemisphere, although the double structure of the southern hemisphere is missing. Related to this, there is an underestimation of the westerly winds for the surface and middle tropospheric between 50°S and 60°S. A major deficiency of the models is the underestimation of the tropospheric mass streamfunction, the quantity that represents the atmospheric overturning of the Hadley Cells and the meridional transport of potential energy. Both the Hadley Cells of the tropics and the indirect cells associated with the Rossby Waves of middle latitudes are too weak.

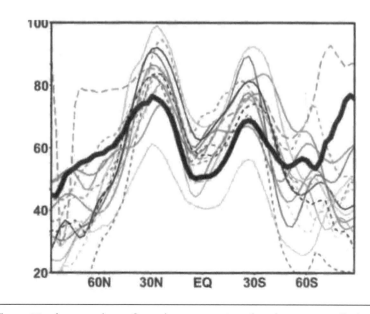

Surface net longwave radiation
Greenhouse gases in the atmosphere interact with the climate system through net long wave radiation at the earth's surface. The doubling of carbon dioxide concentration is expected to reduce net longwave radiation at the surface by about 4 W/m^2.

The computer models compared under the Coupled Model Intercomparison Project (CMIP) tended to overestimate net surface longwave radiation over tropical regions and underestimate the net radiation over polar regions. The differences between models and from estimated values of the climate system were several tens of W/m^2 and significantly larger than expected greenhouse forcing.
(The scale of the graph is W/m^2.)

Figure 25: A comparison of zonal average net surface longwave radiation for different computer models

The representation of surface energy components in the computer models (solar insolation, longwave radiation, eddy heat exchange, latent heat exchange and net heat exchange) vary from model to model and the values are significantly different from estimated values. Although our knowledge about the surface energy components is relatively poor, and the estimates from observations should not be used as a verifying benchmark, the marked difference between models is a profound weakness. These components are vital for regulating the energy flow through the climate system. The very basis of the anthropogenic greenhouse gas scenario is that small changes in surface net longwave radiation will lead to dangerous climate change. If the models do not accurately specify these components then they are unlikely to be representative of what is happening within the climate system. The representations of zonal average net longwave radiation at the surface for the different computer models are shown at Figure 25.

Ocean Heat Transport
The northward transport of heat by the ocean component for each of the computer models of the Coupled Model Intercomparison Project (CMIP), and the estimate (heavy line) for the climate system. (Units: PW = 10^{15} W; positive is northward and negative is southward transport)

On average, the computer models significantly underestimate ocean heat transport and there are significant differences between models (solid line is estimated value).

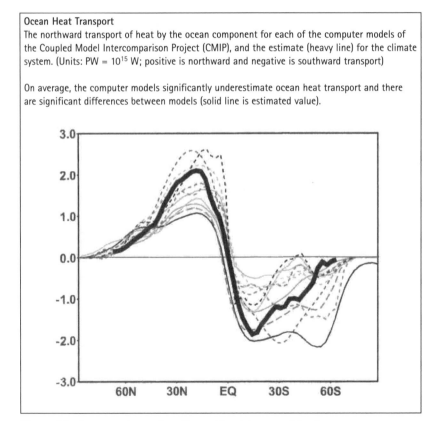

Figure 26: A comparison of northward heat transport by the ocean component of different computer models

One of the difficulties in representing the surface energy components is the large amplitude of the annual cycle, especially over middle and high latitudes of the oceans. During summer there is a large uptake of solar radiation that warms the ocean surface and during winter there is loss to the overlying atmosphere. The net heat exchange over a year is, therefore, the difference between two large quantities. For latent energy exchange and solar insolation the models qualitatively have consistent zonal distributions but the magnitudes vary by 20–30 W/m^2 over parts of the tropics.

As a consequence of the inability to adequately specify the net surface energy exchange at the ocean-atmosphere interface there are deficiencies in the atmospheric and ocean circulations. We have already noted the under-specification of the tropical Hadley Cells of the atmosphere. There is a similar under-specification of meridional overturning in the oceans with inadequate poleward transport of heat. The representations of meridional heat transport by the ocean circulations of different models are shown in Figure 26.

Taking an overview of the representations of the climate system there are systemic deficiencies, and those involving meridional transport of energy by the oceans and atmosphere are particularly relevant. In light of the inter-model differences it is surprising that they are able to maintain relatively similar and stable meridional profiles of near surface air temperature, precipitation and mean sea level pressure. Clearly the computer models are not overly reliant on the meridional transport of energy to maintain the climatological distributions of temperature, pressure and precipitation despite the need for meridional transport of energy to maintain the global radiation balance. The overturning circulations of the atmosphere and oceans are grossly different and consistently underestimated when compared to observations.

ii. CMIP (Phase 2)

In the second phase of the intercomparison, the computer models were forced by a one-percent per year increase in atmospheric carbon dioxide for 80 years. For intercomparison purposes, the differences between the averages over years 61–80 and years 1–20 were taken as the response of the models to the increased greenhouse gas concentrations.

One of the outcomes is that by the time the atmospheric carbon dioxide concentration has doubled (at about year 70) the mean global surface air temperature of the models has only risen by about 2°C. This is significantly less than the average value of 3°C previously obtained for equilibrium at double carbon dioxide concentration. But more surprising is the reduction of variation in the simulated global warming between models. In earlier equilibrium models the variation was a factor of 3 but has been reduced to about 25 percent in the time-evolving models. One explanation for this convergence is the acceptance and adoption of similar formulations for many of the processes that have to be represented in the computer models by different researchers. However, for many of the processes there are few relevant quantitative data against which to tune the formulations.

The projected increase in surface air temperature is greatest over land and over high latitudes of the northern hemisphere. The warming is relatively uniform over the tropical and middle latitude oceans but higher temperature increases are experienced over continental regions. The increase in surface air temperature is reflected in warming throughout the troposphere but there is cooling in the stratosphere that increases with height.

The precipitation responses of the models vary by a factor of 5, from an increase of 0.03 mm/day to 0.015 mm/day. The lower precipitation rate increase is equivalent to additional latent energy released within the troposphere of less than 1 W/m^2 and the higher precipitation rate represents an additional heating rate of the troposphere of more than 4W/m^2, the full radiation forcing from a doubling of carbon dioxide concentration. The area of most precipitation increase is the equatorial western Pacific Ocean but there are also increases over high latitudes of both hemispheres. The computer models show reduction in precipitation over parts of the subtropics of both hemispheres. Consistent with the enhanced hydrological cycle, the computer models produce an increase in specific humidity throughout the troposphere, and the increase is a maximum at the equator near the surface.

Over tropical oceans, the reduction in net longwave loss from the surface that is a consequence of increased concentrations of greenhouse gases can be largely compensated by increased evaporation. A decrease in longwave radiation at the surface of 4 Wm^{-2} (the expected change through a doubling of carbon dioxide concentration) is equivalent to an increase in evaporation of 0.14 mm/day. Latent energy is the fuel of deep tropical convection and any increase in evaporation will enhance the tropical Hadley Cells, the atmospheric angular momentum and the meridional transport of energy. If all of the additional latent heat from evaporation were exported out of the tropics the meridional energy transport would increase by 0.25 PW, or about 5 percent.

The change in net surface heat exchange is generally positive, indicating heat gain by the oceans. The regions of maximum gain are in the North Atlantic and across the Southern Ocean near Antarctica. The reduction in net longwave radiation over most of the tropical and mid-latitude oceans is almost balanced by increase in evaporation and latent energy to the atmosphere.

There is a significant change to the ocean circulation during the enhanced carbon dioxide simulations. The consistent slowing of the ocean overturning is additional to the underestimated ocean overturning and meridional heat transport of the control runs. The reason for the slowing can be traced to the increased heat input to the oceans over high latitudes of the North Atlantic and around Antarctica. These are regions where latent heat flux, sensible heat flux and solar insolation are all reduced as a consequence of increasing the concentration of carbon dioxide. However, the direct longwave radiation forcing over these regions is similar in magnitude to other ocean regions, suggesting that regional feedback processes are being strongly amplified. We should be very cautious about accepting the ocean slowing as being realistic because of the linkage to the poorly specified components of the surface energy exchange.

There is little change in the strength of the tropical Hadley Cells during the enhanced carbon dioxide simulations. The latter is despite the increased tropical evaporation and increased equatorial precipitation observed in the simulations. If deep convection were well specified then the increased latent energy release would be expected to increase atmospheric overturning and the generation of atmospheric angular momentum, for example as happens in the atmosphere during El Niño events. The only increases in zonal winds are in the stratosphere of each hemisphere – extraneous to the changed physical forcing of the atmospheric boundary layer and hydrological cycle of the troposphere. The subdued response to enhanced latent heat release does suggest that the computer models inadequately represent deep convection processes and overly disperse momentum in their representations of sub-grid scale friction processes. This would also help to explain the dispersion of moisture from the models' tropical boundary layer.

The Atmosphere Model Intercomparison Project studies mentioned earlier[151] identified an underestimation of interannual variability of atmospheric angular momentum when computer models of the atmospheric circulation are forced with actual sea surface temperatures. The underestimation of multi-year variability is even more pronounced in long simulations of computer models of the full climate system, both with and without increasing concentrations of greenhouse gases[152]. The smoothed time series of total atmospheric

151 Hide, R et al, 1997. Ibid.
152 Huang, H-P., K.M. Weickmann and C.J. Hsu, 2001. Trend in atmospheric angular momentum in a transient climate change simulation with greenhouse and aerosol forcing. J. Climate. vol 14, pp1525–1534.

angular momentum constructed from observations (the NCEP/NCAR re-analysis data) has an amplitude range of about 1×10^{25} kg m^2 s^{-1} over the period 1958–1998, with an increase during the middle 1970s. In contrast, the variability over a near 200-year computer simulation with fixed atmospheric concentrations of greenhouse gases is only a fraction of this. For a 200-year simulation, where the carbon dioxide concentration in the atmosphere increased three-fold, the increase in atmospheric angular momentum was only 4×10^{25} kg m^2 s^{-1} by the end of the simulation.

The CMIP assessment of the control climates simulated by the coupled models[153] makes points that are pertinent to the enhanced carbon dioxide simulations. The apparent errors in the ocean circulations relate to some combination of erroneous surface energy exchanges and/or transport formulations. If the errors relate to transport they are probably not due to model resolution because some of the biggest differences are between models with high resolution for the ocean component. The correct simulation of the distribution of surface energy exchanges and the associated transport of heat in the ocean is a major challenge in the further development of coupled models.

Simulations of the 20th Century Climate

The apparent ability of computer models to realistically simulate the global surface temperature pattern of the 20th century is one of the principal reasons given by the IPCC for its confidence in the ability of computer models to predict future climate under anthropogenic greenhouse gas forcing. This apparent realism in simulations of the 20th century surface temperature is illusory because the outcomes are dependent on assumptions made about the forcing factors.

The record of global average surface temperature since 1860, as measured by instruments, has significant year-to-year variability. Notwithstanding the short-term fluctuations, an overall trend can be identified that has an increase of about 0.6°C during the 20th century (see Figure 3). The increase occurred over two periods, each with a rise of about 0.4°C; firstly from 1910 to 1945 and then from 1976 to 2000. The intervening interval, 1946 to 1975, exhibited a slightly negative trend. The century scale pattern of global average surface temperature provides a more demanding benchmark for climate models than the comparison with climate statistics at model equilibrium. In assessing climate models against the benchmark of changing surface temperature it is necessary that the simulations not only reproduce the century scale global warming trend but also that they reproduce the three-decade hiatus during the middle decades of the century.

---

153 Lambert, S.J. and G.J. Boer, 2001. Ibid.

There is the possibility that the temperature variability over the 20th century may have been due to internal variability of the climate system. However, the relatively short instrumental record and limitations of reconstructions from proxy data mean that there is only limited data on which to deduce the extent of climate variability on the multi-decadal to centennial timescales. IPCC concludes these limitations "leave few alternatives to using long control simulations with coupled models to estimate the detailed structure of internal climate variability"[154]. In the absence of corroborating data, it is a leap of faith to assume that the lack of internal variability in the computer models is a true reflection of the climate system.

IPCC has compared global surface temperature anomalies from three different computer models that have each performed 1,000-year simulations. No simulation shows a trend in surface air temperature as large as the observed trend over the 1860–2000 instrument period. On the assumption that internal variability is correct in the models, IPCC concludes that recent global warming is not likely to be due to variability produced within the climate system alone[155]. Such a conclusion is inconsistent with the stated belief expressed by IPCC that computer models are an unsatisfactory alternative to actual observations of the climate system. The only valid conclusion that can be drawn is that the computer models do not exhibit internal variability of the scale exhibited by instrument measurements since 1860. However, internal variability of the climate system cannot be discounted as contributing to the observed surface trend.

On the presumption that the computer models adequately reflect the internal variability of the climate system then IPCC has turned to external forcing as the cause of the trend and multi-decadal fluctuation of the surface temperature record since 1860. Processes that might potentially interact with the climate system and cause a shift from equilibrium are those that interact either directly or indirectly with the earth's radiation balance, the so-called radiative forcing. The two dominant natural processes that have been considered potentially important on the century scale are variations in solar irradiance and the impacts from volcanic activity, when large mass of aerosols are injected into the stratosphere. Human activities potentially interact with a number of climate processes to change the radiation balance of the earth. Industrial and transport processes disperse greenhouse gases and aerosols into the troposphere while agricultural, pastoral and other land management activities can modify natural greenhouse gas exchanges, disperse mineral and carbon-based aerosols, and change the surface albedo (reflectance).

---

154 IPCC, 2001, Ibid. p702.
155 IPCC, 2001, Ibid. p703

The nature of natural and anthropogenic processes that potentially interact with the earth's energy balance are well understood and climate modellers have applied seemingly plausible forcing during simulations of the 20th century climate. Estimates of solar irradiance variability and volcanic activity are considered to be the main natural forcing processes, and increasing greenhouse gas concentrations and varying sulphate aerosol concentrations in the atmosphere are the sources of anthropogenic forcing. Although the processes and how they interact with the earth's radiation are qualitatively understood, the magnitude of the interactions is, in many cases, poorly quantified.

i. Solar radiation

Satellite measurements of solar irradiance have only been made since 1978, a period covering two solar cycles. As previously noted,[156] fluctuations of solar irradiance, and hence changing solar forcing of the climate system, have been linked to changing sunspot numbers. However, there is not a direct link between sunspot numbers and solar irradiance. Over the period of satellite measurements the number of sunspots observed over a cycle ranges from a minimum number near 10 to a maximum number of more than 100 but the corresponding range of solar irradiance is only about 1.5 W/m^2. The estimation of earlier solar irradiance, although based on physical mechanisms, is indirect. Reconstructions of solar irradiance since 1750 used in different computer models vary in amplitude and phase.

Observed sunspot numbers would suggest that solar irradiance increased from a minimum about 1900 to the middle of the century before stabilising at near the current value. That is, there was solar forcing in the early half of the century but none over the later half. However, not only are there uncertainties pertaining to the relationship between sunspot number and solar irradiance, there are also conflicting views on the correct reconstruction of solar irradiance over the past two decades of satellite observations. A recent reappraisal of the solar irradiance analysis from satellite measurements suggests that there may have been an increase in solar irradiance over the past two cycles[157].

One potential form of solar forcing that is not included in computer models is the potential impact of cosmic rays on cloudiness, and hence earth's albedo. The cosmic ray-cloudiness linkage has been speculated to be very significant based on linkages with the 100 million year modulation of glaciation but there is no data or theoretical basis to include a quantitative forcing due to the relationship.

156 See Section, Variations of solar irradiance.
157 Wilson, R.C. and A.V. Mordvinov, 2003. Ibid.

ii. Vulcanism

Volcanic eruptions inject enormous quantities of fine particles into the stratosphere and the zonal winds quickly spread the aerosols around the hemisphere to form a continuous zonal band. With time the fine particles spread and form a layer of global extent. The volcanic ash enhances the reflection of solar radiation to space (cooling) and it also interacts with absorption and emission of longwave radiation. Satellite-based observations during the Mt Pinatubo eruption of 1991 clearly show, over the tropics, the reduction in longwave thermal emission, the increase in reflected solar shortwave radiation, and the overall decrease in net radiation to earth[158]. The reduction in net radiation had a peak value of near 9 W/m^2 but was relatively short-lived, being indistinguishable after 2 years.

Vulcanism is a natural external forcing of the climate system and the satellite observations during the Mt Pinatubo eruption provide a quantitative estimate of the magnitude and duration of that event. The satellite observations relating to the Mt Pinatubo eruption suggest that volcanic ashes settle out of the stratosphere relatively quickly and there is negligible effect after about two years. Nevertheless, if there is a sequence of volcanoes the recurring injection of ash will likely effect the global temperature and it is appropriate to include this external forcing in computer models that simulate past climate.

During the period of satellite observations since 1979 there have been two major eruptions that have affected the satellite signal; El Chichon in 1982 and Mt Pinatubo in 1991. The impacts on the earth's radiation by these two events were different, in part reflecting their different latitudes of location and in part reflecting different intensities. This points to uncertainty in reconstructing the radiative forcing during the 20th century due to vulcanism. Volcanoes impose a negative radiation forcing on the climate and it is estimated that 20th century activity was greatest before 1915 and after 1960. More recent representations of volcanic ash forcing of the climate for the 20th century provide individual latitude-intensity patterns for each event[159].

iii. Aerosols

Tropospheric aerosols are the fine particulates, often clearly observed as haze, in the lower layers of the atmosphere. There are many sources, including lifting of dust from arid or desertified land, burning of natural vegetation (both land clearing and wildfires), urban and

158 Wielicki, B.A. et al, 2002. Ibid
159 Ammann, C.M., G.A. Meehl and W.M. Washington, 2003. A monthly and latitudinally varying volcanic forcing dataset in simulations of 20th century climate. Geophys. Res. Lett., vol 30, No.12 (59)

industrial combustion, and emissions as by-products of industrial processes. The ocean is a significant contributor to the atmospheric aerosol loading because of the lifting and evaporation of sea spray leaving suspended marine micro-organisms. Natural dust concentrations in the atmosphere vary on seasonal and longer timescales, particularly over arid and semi-arid regions. Dust loading varies with the seasonal rainfall patterns and increases significantly during prolonged drought. Some models attempt a drought index as a basis for estimating natural variations in atmospheric aerosol loading but there are only few quantitative observations to support these refinements.

Aerosol loading and variability in the troposphere, including that due to human activities such as land-use, combustion and resource processing, are difficult to quantify. Some broad estimates of persisting atmospheric concentrations of aerosols have been made based on air quality and other measurements. Analysis of the physics and chemistry of aerosols can broadly distinguish between those occurring naturally and those resulting from human activity but this is of little help in estimating the magnitude of aerosol forcing of the climate system.

The interaction between different aerosol types and radiation is poorly understood. The impact of sulphates and biomass burning is claimed to contribute to cooling while black carbon from fossil fuel burning might produce warming or cooling. Estimates of aerosol radiative forcing used in computer models have not been derived from the physics and chemistry of aerosols but under the premise that the observed warming during the industrial era has been due to human activities[160]. These estimates of aerosol radiative forcing are, at best, speculative. Indeed, IPCC cautions that simulations including estimates of indirect sulphate forcing should be regarded as preliminary[161].

iv. Anthropogenic greenhouse gases

Radiation transfer through the atmosphere can be calculated with accuracy under conditions of clear skies, known temperature and moisture profiles, and known atmospheric carbon dioxide concentration. These calculations have been established from a long history of laboratory experiments and validated by observations at the surface, from aircraft flying at various altitudes and from satellites. Even under conditions of changed carbon dioxide concentrations, the laboratory observations give confidence to the predicted radiation

160 Anderson, T.L., R.J. Charlson, S.E. Schwartz, R. Knutti, O. Boucher, H. Rodhe and J. Heintzenberg., 2003. Climate forcing by aerosols – a hazy picture. Science, vol 300, pp1103–1104.
161 IPCC, 2001. Ibid p711.

transfer values. However, the introduction of clouds into the calculations markedly increases uncertainty to the outcomes. It is not possible to specify the cloud dimensions and microphysical properties in a sufficiently simple way to represent the range of radiation interactions observed in the troposphere.

The change to longwave radiation resulting from increasing greenhouse gas concentrations in the atmosphere is important in three aspects of the global energy budget:

(i)   the reduction of longwave radiation emitted to space at the top of the atmosphere increases the total energy of the global climate system;

(ii)   the increase in downward longwave radiation at the surface reduces the net longwave emission at the surface and energy is retained in the underlying surface;

(iii) the reduction in longwave emission at the top of the atmosphere (effectively the tropopause) and the reduction of net longwave radiation at the surface means that there may be little change to the ongoing radiation cooling of the troposphere.

Overall, the primary impact of increasing atmospheric greenhouse gas concentrations is not felt in the troposphere (where strong radiation cooling continues) but as a retention of energy in the underlying surface. We have already noted that net longwave radiation at the surface is poorly specified, with significant differences between computer models and significant latitudinal biases in all models[162]. The variance between different computer models exceeds 25 W/m^2 in the tropics, it exceeds 30 W/m^2 over the Arctic and it exceeds 60 W/m^2 over the Antarctic. These are very big numbers when compared to the estimated 2 W/m^2 increase due to increasing concentrations of atmospheric greenhouse gases during the 20th century and 4 W/m^2 for a doubling of CO_2 since industrialisation. The models systematically overestimate surface net longwave radiation in the tropics and underestimate longwave radiation over polar latitudes. These errors in the climate 'drivers' must necessarily be offset elsewhere in the models by other process errors because the computer models are internally consistent and, under equilibrium conditions, tend to a realistic statistical representation of the surface temperature, pressure and precipitation characteristics when compared to the observed climate.

---

162 Covey, C., et al 2000. Ibid.

The four forcing factors (solar irradiance, episodes of volcanic activity, the increasing atmospheric loading of aerosols and increasing concentrations of greenhouse gases) have been incorporated in computer models to simulate the climate of the 20th century. The IPCC advocates that the close representation of the simulated global surface temperature pattern to the observed values is indicative of how realistically the models represent the processes of the climate system. The IPCC evidence is illusory because of the uncertainties associated with the representation of the radiation forcing for each of the natural and anthropogenic factors. Equally plausible temporal distributions of the magnitudes of each of the forcing factors could have been adopted with quite different outcomes.

IPCC notes that the computer reconstruction of the global surface temperature for the 20th century relies upon the cumulative impacts

Computer model reconstruction of the global average surface temperature, as presented by the IPCC, to show a) the impact of specified natural forcing (solar radiation and volcanic activity), b) anthropogenic forcing (greenhouse gases and aerosols), and c) the cumulative impact of natural and anthropogenic forcing. The reconstructions are compared with observations.

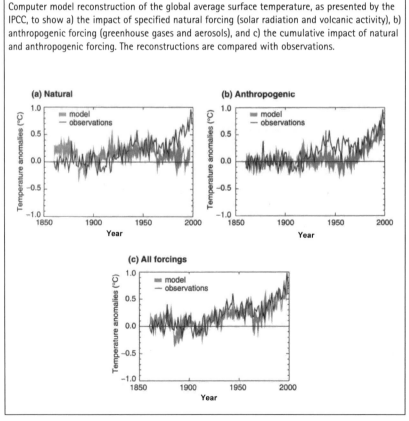

Figure 27: Simulation of the climate of the 20th century with natural and anthropogenic forcing

of the separate natural and anthropogenic forcing factors[163]. However, as shown in Figure 26, neither factor alone reproduces the observed pattern with fidelity. The simulations using only the natural forcing (solar radiation and vulcanism) has a period of cooling lasting several decades about 1900 and cooling after 1950, both events are attributed to volcanic activity and there is no discernible impact of solar forcing. In contrast, the temperature pattern simulated with the anthropogenic forcing has no warming until after 1960, at which time a strong warming trend due to increasing concentrations of greenhouse gases is established.

The global surface temperature pattern produced by the computer simulation with combined natural and anthropogenic forcing is, in broad terms, similar to the pattern that is observed. There are obvious questions about why the early 20th century warming event was shorter and more intense in the combined forcing simulation than in the simulation with only natural factors; and why the middle century temperature decline was delayed in the combined simulation but not when only natural forcing was included. The climate system is complex with interactions between the various processes and we should not expect the outcome to be a linear combination of the individual forcing. It is not clear as to whether the outcome is a realistic representation of the natural and anthropogenic forcing factors or whether serendipity and judicious choice of the time histories of the individual forcing factors are contributing factors!

The apparent ability to simulate the global surface temperature pattern of the 20th century is achieved despite some fundamental uncertainties being unresolved. The limited internal variability of the computer models is no guide to the internal variability of the climate system. In this context, the so-called climate shift in the tropical Pacific Ocean around 1976[164] is an enigma. The warming of the tropical Pacific Ocean is linked to the increase in ENSO activity, the increased circulation of the Hadley Cells and the export of energy from the tropics. The sudden warming of the tropical Pacific Ocean during 1976 is not explained by increased greenhouse gas concentrations in the atmosphere and the sudden ocean warming is physically linked to the global warming that takes place over the next two and a half decades.

There is good reason to believe that the contribution to climate forcing from changing solar irradiance has been underestimated in the computer simulations of the 20th century. In Figure 27a, at the beginning of the simulation of global surface temperature with natural

163 IPCC, 2001. Ibid, p11 and p710.
164 IPCC, 2001. Ibid. p 454.

forcing, it can be observed that there is a positive bias and this is not exceeded during the course of the simulation. There is some uncertainty in the change in solar irradiance over recent decades but reconstructions consistently indicate that values were relatively steady over the second half of the 19th century before increasing strongly through the first half of the 20th century. The simulation with natural forcing does not reflect a solar signal with these characteristics. Rather, the simulation reflects the characteristics of forcing from volcanic activity that reduces the global temperature from the initial state around the turn of the century and again later in the century, as prescribed.

The IPCC claims that the limited internal variability of computer models is realistic. The limited internal variability is only consistent with IPCC's favoured reconstruction of global surface temperature over the past millennium, but the latter is demonstrably flawed. The computer models are highly constrained to maintain a stable simulation and, as a consequence, their internal variability is also constrained. The computer models are primarily validated against static quantities, such as meridional profiles. When compared to observations relating to the more dynamic characteristics of the climate system, such as atmospheric angular momentum, and the energy transports by the oceans and atmosphere, the statistics validate poorly. The sudden warming of the tropical Pacific Ocean during 1976 points is a sudden unexplained climate shift with prolonged global impact. Lack of data precludes classifying the shift as unusual or a symptom of recurring internal processes of the climate system.

The IPCC has represented the impact of increasing greenhouse gas concentrations on the climate system as a radiation problem. In the complexity of the climate system this is an inadequate framework. A more inclusive framework relates how local changes in radiation, as a consequence of increasing greenhouse gas concentrations, interact with the various energy storages and thermodynamics of the global climate system. These interactions will include changing ocean circulations and their impact on the heat storage of the ocean surface layer, growth or contraction of the ice sheets and changes to the rate of meridional transport of energy.

The apparent ability of computer models to simulate the global surface temperature pattern of the 20th century comes with too many assumptions and shortcomings. Despite the IPCC advocacy, it is not possible to isolate anthropogenic greenhouse gases as the cause (or even a major cause) for the observed warming over the last two and a half decades of the 20th century. The worldwide advance of mountain glaciers until the mid-19th century, and their steady retreat since, point toward large-scale natural processes systematically affecting the climate system over prolonged intervals. Whether the systematic processes are internal to the climate system, an outcome of external forcing, or a combination of these cannot be determined with any confidence from existing data and analysis tools. As a corollary, the sensitivity of the earth's temperature response to greenhouse gas forcing cannot be scaled by reference to the magnitude of recent global temperature increase and the forcing by anthropogenic greenhouse gases as represented in computer model simulations of the 20th century.

We have noted that, because of the ongoing net radiation deficit over polar regions, meridional transport of energy is required to prevent continuous expansion of the polar ice sheets. In the same way, changes in meridional energy transport are fundamental to the growth or contraction of ice sheets, and to temperature changes over middle and high latitudes. The characteristics of recent 'global warming' are consistent with the observed higher values of sea surface temperatures over the tropical Pacific Ocean and enhanced meridional transport of energy since 1976. Mountain glaciers have retreated; near-surface air temperatures over middle and high latitudes have risen; and polar ice sheets have contracted, at least for Arctic sea ice. Maximum sea surface temperatures over the western Pacific Ocean warm pool have not risen and nor have tropospheric temperatures. If the late 20th century warming is to be attributed to increased concentrations of greenhouse gases then it is necessary to demonstrate a direct linkage between radiation forcing and the sudden rise of sea surface temperature over the tropical Pacific Ocean in 1976. This has not been done.

Future Climate

More than six years after the December 1997 grand jamboree at Kyoto, when government representatives agreed that internationally sanctioned reductions in carbon dioxide emissions should be adopted by a selected list of industrialised countries, the Kyoto Protocol has not yet come into force. The reason is because, of the listed countries that were required to limit carbon dioxide emissions, a sufficient number of countries necessary to meet the technical requirements of the Protocol had not ratified and lodged the appropriate instruments with the United Nations. The reasons for not ratifying varied between countries but they mainly relate to the envisaged economic impacts. The imposts on industrialised countries for complying, particularly by those countries that do not have a significant nuclear energy base, will increase energy costs and they become less competitive in trade. There is also a view that the Protocol will not be effective unless all countries are committed to constraining greenhouse gas emissions and contribute to meeting the objective of stabilising concentrations of greenhouse gases in the atmosphere.

There are also many sceptics about the science underpinning the UN Framework Convention on Climate Change and its Kyoto Protocol. Criticism has been levelled at various assumptions and conclusions of the IPCC assessment reports but this scepticism has not yet penetrated into any formal government positions. The IPCC continues to be the intergovernmental forum for evaluating climate science and formulating recommendations for related policies at the international level. IPCC issued its Third Assessment Report in 2001 and preparation has commenced in anticipation of a fourth assessment about 2007.

The IPCC First Assessment Report of 1990, although prepared in haste, was cautious. The report confirmed that there is a natural greenhouse effect keeping the earth warmer than it would otherwise be, and that emissions resulting from human activities are substantially increasing the atmospheric concentrations of greenhouse gases. In 1990, IPCC predicted that global average surface temperature would rise 3°C above 1990 values by 2100 and that sea level would rise by 65 cm. In this first assessment, IPCC emphasised that there were many uncertainties in its predictions, particularly with regard to the timing, magnitude and regional patterns of climate change. The 1990 assessment drew attention to areas of uncertainty that could have an impact on its predictions. These included:

- sources and sinks of greenhouse gases, which affect predictions of future concentrations;

- clouds, which strongly influence the magnitude of climate change;

- oceans, which affect the timing of future climate change; and

- polar ice sheets, which affect predictions of sea level rise.

The 2001 IPCC Third Assessment Report is stronger in its language and expresses confidence in its predictions of global warming. The Third Assessment Report suggests that our knowledge of climate and ability to project future climate has advanced considerably since its first report was issued. The Third Assessment Report:

- confirmed that concentrations of greenhouse gases in the atmosphere were increasing and would continue to increase according to a range of scenarios relating to economic development and emission control compliances;

- expressed the view that the observed increase in global surface temperature of the past 50 years was attributable to human activities; and

- expressed confidence that the ability of computer models to predict future climate had increased, and projected that the globally averaged surface temperature would increase in the range of 1.4°C to 5.8°C between 1990 and 2100 (other climate changes were also projected, including for precipitation, sea level rise and storm frequency).

In reviewing the changes in science and scientific technology over the past decade it is clear that the confidence of the IPCC Third Assessment Report is misplaced. There have been advances in all areas of climate science but these are not sufficient to give additional confidence to projections[165] of future climate, nor are they sufficient to attribute recent change to human activities. Rather, the increased confidence is a delusion resulting from a lack of critical analysis of the claimed scientific advances and computer model developments.

Misleading climate projections

The one-dimensional energy budget model of the climate system used by IPCC (Figure 1) to explain its concept of radiative forcing is inadequate. The model gives prominence to radiation processes and the assumptions disguise a range of non-radiative processes that are important components of the climate system. By considering only global and annual average statistics and assuming energy balance, the thermodynamics of the ocean and atmospheric circulations and the processes that regulate the ocean-atmosphere energy exchange are ignored. The IPCC representation is an unrealistic model of the climate system because it portrays the earth as if it were flat and motionless with no pole to equator temperature gradients.

The circulations of the oceans and the atmosphere, the energy reservoir of the tropical ocean surface, the ocean-atmosphere energy exchange and the hydrological cycle (including the varying mass of polar ice) are integral components of the climate system and cannot be ignored. Further, the interactions between the components are highly non-linear. The climate system does not respond in a linear manner to imposed changes, whether they be direct changes to solar irradiance or indirect changes, including through changes in greenhouse gas concentration, aerosol loading or surface albedo. To represent the climate system as a one-dimensional pseudo-linear model is highly misleading. Such a model can only explain in the most rudimentary and qualitative way how the earth's climate might respond to small natural or anthropogenic imposed changes.

The only useful characteristic of the one-dimensional energy budget model is that it does clearly show the relative magnitudes of the various terrestrial radiation components of the climate system. Of particular importance, it shows that the troposphere continually loses energy through radiation processes. Convection within the troposphere, including the need for convective instability, is the crucial process that regulates the distribution of energy from the atmospheric boundary layer to the troposphere, but the one-dimensional model cannot show this. The inadequate representation of convection has important ramifications for the interpretation of

---

165 IPCC has made a subtle change of language between assessments. In the First Assessment Report, IPCC made 'predictions' whereas in the Third Assessment Report it was careful to refer to future climate states as 'projections'. The reasons for this change are neither obvious nor explained.

climate response including that changes in radiation will lead to changes in the temperature of the troposphere. Where the temperature of the troposphere directly affects the magnitudes of terrestrial radiation the reverse is not true. Terrestrial radiation and its changes can only affect tropospheric temperature indirectly by changing the surface temperature.

An assumption of the IPCC model is that there is energy balance at the earth's surface. This is not true because the ocean surface layer, in particular, has large thermal capacity that is not recognised in the simple one-dimensional energy budget model. Solar radiation penetrates to depth in the ocean surface layer, which is effectively an energy reservoir. Sea surface temperature is a primary control on the exchange of heat and latent energy between the ocean and atmosphere and is a function of a range of other processes in addition to radiation.

The simple one-dimensional energy budget model does not give any information about long-term exchange of energy between the earth's climate system and space during the expansion and contraction of polar ice sheets. Energy is extracted from the climate system during snow formation. This is because the latent energy taken up from the oceans by water vapour during evaporation is less than that given up to the atmosphere (and ultimately to space) as the water vapour forms snow particles. During expansion of polar ice there is a net loss of energy from the climate system to space as water is transferred (via evaporation and then snow formation) to the polar ice sheets. At the top of the atmosphere the emission of terrestrial radiation to space exceeds the absorbed solar radiation. Similarly, during contraction of polar ice sheets, solar radiation absorbed by the climate system exceeds the emission of terrestrial radiation.

It needs to be emphasised that a global imbalance of solar and terrestrial radiation at the top of the atmosphere does not necessarily lead to expansion or contraction of the polar ice sheets. Over the polar regions, at mid-summer, there is always an excess solar radiation over terrestrial radiation and the availability of energy to melt polar ice is not an issue. The ice mass is therefore very stable when temperatures are below 0°C, irrespective of the availability of energy from solar radiation or meridional transport.

The most serious deficiency of the one-dimensional energy budget model is that it masks the fundamental variation of solar radiation between the tropics and the poles. The excess solar radiation over the tropics and the deficit over polar regions means that the atmosphere and ocean circulations, and the thermodynamic processes associated with motion on a rotating earth, are crucial as the climate system is continually adjusting towards a radiation balance. Radiation balance, when achieved, is only transitory and not a stable climate state. The seasonal cycle of surface temperatures over middle and high latitudes are a reflection of enormous energy transports as polar regions oscillate between slight solar energy excess at midsummer and the deficits during the long dark of winter. The relatively small anomalies induced in the terrestrial radiation fields become almost indistinguishable as the excess solar radiation absorbed in the energy reservoir of the tropical oceans is exchanged and transported through the climate system before its ultimate emission to space over polar regions.

IPCC advances the erroneous view that the climate system has been stable with near constant global surface temperature for the past millennium, and that recent warming is unusual. This is a view that is at odds with many individual proxies, albeit proxies without 'annual resolution' and hence not useful for multi-proxy reconstruction. Many proxies from widely varying locations, including mountain glaciers, are consistent in their support for a picture of global warmth during the Medieval Period, a cooling during a Little Ice Age, and a more recent recovery of warmth. The multi-proxy reconstructions of global surface temperature over the last 1,000 years are based on inadequate data, they lack fidelity and they are misleading. Not only are the proxies inadequate in terms of the fidelity of annual temperature variability for their local region but the number of proxies and their locations are not representative of the hemispheric scale to which the reconstructions are ascribed.

To give further support of its overall view that the climate system is relatively stable, unless forced by an external factor, the IPCC claims that the climate system has only limited internal variability. This conclusion is deduced entirely from the behaviour of computer models and is erroneous. The conclusion ignores the rudimentary stage of development of computer models and it is only in the most recent computer models that air-sea exchanges of energy have not been constrained by predetermined adjustments. Until these recent refinements, for which the benchmark judgement of success is the stability of model simulations, it had been necessary to prescribe ocean–atmosphere exchange parameters to ensure that process errors did not propagate and prevent the achievement of an equilibrium state. The atmospheric component of the models, which has the longest history of development, still lacks responsiveness to interannual forcing. The ocean component cannot be similarly tested because there is no benchmark data on variability of the ocean circulation.

The computer models are an inappropriate benchmark for assessing the degree of internal variability within the climate systems. The atmosphere is very responsive to changes in sea surface temperature patterns and this is clearly shown during El Niño events when the atmospheric angular momentum increases and additional energy is transported poleward. This observed variability during ENSO is not reproduced in the forced atmospheric models. The ocean circulations are relatively slow moving but they are forced by wind stress at the surface. There is observed variability of the surface and sub-surface temperature characteristics on decadal and longer timescales. There is also evidence that slow changes to the meridional overturning of the oceans (the thermohaline circulation) are occurring. These slow changes would be expected to modulate surface climate, especially through changes that feed back to the meridional transport of energy by the atmosphere. There are no observations against which to quantify the decadal and longer period variability of the oceans, but this is no reason to ignore such variability in climate models and projections of future climate.

The sensitivity of climate to variations in meridional transport of energy is gauged by the magnitude of the energy transported poleward annually. The annual mean transport toward the poles in each hemisphere has been estimated, based on observations, as about 5 PW. To put this in perspective, an increase of one percent persisting for less than 10 years would transport enough additional energy to the Arctic to melt the existing sea ice volume. The tropical oceans represent an enormous energy reservoir and variations in the rate of energy exchange between the oceans and atmosphere will feed directly into changes in the temperature and ice mass over the middle and high latitudes. The observed increase in atmospheric angular momentum provides supporting evidence that northern hemisphere warming of recent decades has been a direct consequence of the climate shift to warmer temperatures over the tropical Pacific Ocean in 1976.

The IPCC relies on computer models for the veracity of its projections of global surface temperate through the 21st century. To the casual observer the equilibrium climate of these models has many similarities to observations. The simulations for the 20th century, using seemingly plausible natural and anthropogenic forcing, reproduce characteristics of the observed record of global surface temperature. These characteristics of the computer model might suggest that they are replicating the processes of the climate system with a degree of fidelity and should provide useful projections under well-specified forcing scenarios. However, an examination of those characteristics of the computer models that relate to the interaction between terrestrial radiation and the climate system, and that relate to the meridional energy transport of the computer models, reveals chronic deficiencies. It is difficult to decide whether the computer models were able to reproduce the superficial correspondence because of serendipity or because of judicious choices of required parameters and forcing functions.

A comparison has been made between equilibrium characteristics of computer models and observations (or best estimates from limited observations) as part of the Coupled Model Intercomparison Project (CMIP). For the purposes of identifying the impact of changing greenhouse gas concentrations in the atmosphere, the most important characteristic of the climate system is the net longwave flux at the earth's surface. It is the changes in net longwave radiation at the surface, forced by greenhouse gases, that modulate surface temperature and hence the energy flow through the climate system. Unfortunately there is no consistency of this quantity between the different computer models. Clearly some of the models must be grossly in error but it is not possible, because of the limited number of observations contributing to the benchmark, to determine which model might be nearly correct and which to discard as grossly misleading.

There are also significant differences between the computer models in their ability to reproduce atmospheric angular momentum and mass overturning in the tropical Hadley Cells. There is significant variance between models in the speed of the subtropical jetstreams of the high troposphere and in the intensity of mass overturning in the Hadley Cells. This aspect of the meridional transport of energy is clearly being underestimated. From other studies it has been shown that atmospheric angular momentum in the computer models is not as responsive to interannual forcing as the earth's atmosphere. The intercomparison of computer model projections, after a century of forcing by one percent per year increase in greenhouse gas concentrations, shows very little increase in the intensity of Hadley Cell circulation. The computer models, under greenhouse gas forcing, certainly do not project the significant increase in Hadley Cell circulation and atmospheric angular momentum that has been observed in the atmosphere since the climate shift of 1976.

The inadequate replication of the ocean component of the computer models is gauged from the characteristics of ocean heat transport. Not only do the models exhibit large differences over the important middle latitudes of both hemispheres but also they consistently underestimate the magnitude of the transport.

These few examples of characteristics of the computer models relevant to radiation forcing and meridional energy transport starkly highlight inadequacies. The computer models do not adequately represent net longwave radiation at the surface and, as a consequence, they are not representing with any fidelity the fundamental interaction between terrestrial radiation and the broader climate system. Changes in net radiation at the surface due to a doubling of atmospheric carbon dioxide concentration are significantly less than the variability between models and the average difference between the models and the accepted benchmark. Similarly, the computer models are underestimating the fundamental meridional transports of energy necessary to achieve global net radiation balance of the climate system.

IPCC's claim, that computer model simulations that include estimates of natural and anthropogenic forcing reproduce the large-scale changes in surface temperature for the 20th century, must be treated with caution. The prescribed effects of volcanic activity dominate the natural forcing response. There is no discernible effect of increasing solar activity that is generally acknowledged to have taken place over the late 19th and early 20th centuries. Whether the increasing solar activity continued into the late 20th century is controversial. Clearly the method of representation of solar radiation in the computer model simulation affects the sensitivity – a top of the atmosphere change to net radiation, as under the IPCC radiative forcing hypothesis, is much less sensitive than direct solar energy input to the ocean surface layer and land surface.

Overall, the assessment of the computer models by IPCC is very generous. The assessment does not adequately reflect deficiencies that severely limit the ability of the computer models to simulate the climate system and project, with veracity, future climate under a scenario of increasing atmospheric greenhouse gas concentrations. The IPCC focus on equilibrium statistics ignores the underlying deficiencies of the thermodynamics and energy transport. IPCC fails to acknowledge that the 'global warming' projected by the computer models might be as much due to propagation of disguised process errors as due to a genuine response of the modelled climate system to greenhouse forcing. The IPCC confidence in the performance of computer models is misplaced and, like the simple one-dimensional model, the computer models are misleading and lead to false projections.

Planning for an uncertain world

The 1985 Villach Conference, that launched anthropogenic climate change as an international policy issue, was fortuitous in its timing. The 1982–83 El Niño was, at that time, the most intense of the century, and its worldwide economic, social and environmental impacts had received widespread publicity. The impacts of the event were particularly severe across tropical and subtropical countries but the impacts also extended to the middle latitudes. The 1957–58 International Geophysical Year and its coordinated programs of research had initiated studies that linked the atmosphere and oceans and their effect on climate. The 1982–83 El Niño event gave further emphasis to climate being an integrated system rather than a set of seemingly unrelated geographic happenings, especially of droughts and floods. Measurements of atmospheric carbon dioxide and other greenhouse gases that had commenced with the 1957–58 International Geophysical Year were recording a systematic trend with increasing concentrations.

By 1985 the global community was sensitised to the reality of climate events wreaking death and destruction, especially in developing countries with only limited resilience. The media reports and images from 1982–83 portrayed drought, desertification and famine on a wide scale, floods, tropical cyclones and coastal inundation from storm surges. The added ingredients of a simple theory for global warming and demonstrably increasing concentrations of carbon dioxide and other greenhouse gases in the atmosphere were sufficient to trigger a widespread response for global action. Unfortunately, the road to Kyoto was a pathway that was not going to achieve beneficial outcomes for the global community. The actions proposed to reduce anthropogenic climate change will not see any reduction to the devastating impacts of climate extremes such as were witnessed during the 1982–83 El Niño event, and much the same later during the major 1997–98 event.

Also included in the 1985 Villach Conference Statement, but largely ignored, was a general recognition that planning and the development of mitigation and response strategies could make communities more resilient and avert many of the impacts from climate extremes. The cautionary note was that the climate statistics gathered over the past century were only a limited guide for the climate of the future. It was here that the warning about anthropogenic climate change was introduced and diverted the message. The international response, and the road to Kyoto, has been to confront the perceived threat of anthropogenic climate change, while the need to plan for climate extremes and potential change has been largely ignored, especially the needs of people in less developed countries.

It is now timely to look again at natural climate variability and the extreme events on the interannual to decadal timescale that have and will continue to plague the global community. ENSO and prolonged climatic impacts, such as the 1930s North American Dust Bowl and the Sahel drought of the 1960s and 1970s, need to be better understood if environmental degradation, property destruction and community hardship are to be avoided and food security enhanced. There is also the insidious drying of many subtropical continental areas that has occurred over the past 5,000 years and may be still continuing as ocean temperatures slowly warm. There is no reason to be confident that in the near future the earth will not warm to similar temperatures achieved during past interglacials.

There is a need to build both a better understanding of past variations of climate and improved predictive tools if we are going to better prepare for the future. Unfortunately, in the rush to get worldwide action on greenhouse gas emissions, there has been a tendency to place too much reliance on the predictions of computer models and insufficient credence to data relating to past climates. Similarly, the concept of radiation forcing of the climate system, as advanced by the IPCC and central to much of the computer modelling, is an incomplete framework for identifying the likely scope of future climate change. The limited framework of anthropogenic climate change presupposes that the natural climate system is stable or is only varying because of the natural radiation forcing components, such as by changing solar irradiance or volcanic aerosols.

In reality, a major factor regulating global climate is the ongoing transport of energy from the tropics to polar regions by the motions of the atmosphere and the oceans. More vigorous circulations transport more energy poleward to warm surface temperatures and melt polar ice. The atmosphere is responsible for nearly 90 percent of the annual transport in each hemisphere and varies seasonally with the annual cycle of surface heating and cooling over the polar regions of each hemisphere. The oceans, while contributing little more than 10 percent of the annual meridional energy transport, are primary reservoirs for absorbed solar radiation. The build-up of heat in the ocean surface layers over the tropics is the source of heat and latent energy to the atmospheric boundary layer. The patterns of wind and sea surface temperature regulate the exchange of energy between the tropical oceans and the atmosphere and, indirectly, provide a control over the meridional transport of energy by the atmosphere. Warmer tropical oceans exchange more energy with the atmospheric boundary layer than cooler waters.

The ocean currents have relatively large thermal capacity and mass inertia and vary only slowly with time. Even the relatively fast changes in sea surface temperature patterns across the equatorial central and eastern Pacific Ocean in conjunction with El Niño events are on the interannual timescale. The wind-driven surface gyres of the subtropical oceans apparently vary over decadal and longer timescales. Information about past changes in the thermohaline circulation, or ocean overturning, is qualitative but suggests that variations occur on the centennial and longer timescales. There are only limited linkages between the water bodies of the various ocean basins but they all interact with the atmospheric circulation. The response of the atmosphere to the thermal forcing of an individual ocean basin will be relayed to the circulations of all the ocean basins.

The underlying variability of the ocean circulation and the responsiveness of the atmosphere to sea surface temperature patterns, especially those of the tropics, identify processes for significant internal variability of the climate system. One of the outcomes from internal variability will be fluctuation in the poleward transport of energy by the atmospheric circulation. There is evidence of increased poleward transport of energy occurring during the episodic El Niño events. There is also evidence that the warmer tropical sea surface temperatures of the tropical Pacific Ocean since 1976 have also contributed to a more active atmospheric circulation and enhanced transport of energy to middle and higher latitudes during much of the past three decades. The transport of additional energy to middle and high latitudes since 1976 is a source for the observed 'global warming' and melting of polar ice and mountain glaciers.

The arguments that recent global warming and the depletion of land ice volumes is other than part of a pattern that became established in the mid-19th century are not compelling. The rate of warming over the later decades of the 20th century is not very different from what occurred over the early decades. The pattern of warming and ice depletion is a reversal of the pattern of the previous 400 years, during which there was cooling, expansion of polar sea ice (at least for the Arctic) and advance of mountain glaciers. Available cultural and palaeoclimate data suggests that this is a pattern that had been recurring over several thousand years.

The impact of human activity is yet to be established in the context of change to the climate system. Radiative forcing, whether by natural or anthropogenic causes, is only linked to climate change through the use of computer models. There continues to be imponderables about the extent of natural forcing through the influence of variations in solar irradiance and galactic cosmic rays. There is tantalising evidence that these latter are effective in moderating climate but the theories still lack fully developed and evaluated process descriptions. Also, the necessary substantial body of observations for validation is incomplete.

At their current stage of development, computer models of the climate system are constrained by the need to maintain a stable 'climatology' that approximates what currently prevails on earth. In the absence of extensive data, their development is also constrained by perceptions about how processes of the climate system should respond over time. As a consequence, computer models are an agent for a self-fulfilling prophecy – their projections accord with prior expectations.

These weaknesses are clearly demonstrated by early simulations for the prediction of climate change that utilised an ocean slab construct to manage the ocean-atmosphere exchanges of heat, moisture and momentum. In such a formulation the change in model outcome can only result from predetermined processes, including the applied radiation forcing. The more sophisticated computer models currently in use project similar global warming to those crude representations of earlier times.

More recent formulations of the computer models do attempt a dynamic representation of the circulations of the oceans but the overturning and poleward mass and heat transport are consistently underestimated. Despite this major inadequacy the simulations are able to reproduce surface air temperature, surface air pressure and, to a lesser degree, precipitation fields similar to prevailing climatology. Three areas of uncertainty in the model formulations are: in the processes associated with ice-melt at the margins of ice sheets; the specification of clouds and their interaction with radiation; and in the formulation of longwave radiation at the earth-atmosphere interface. For each of these processes there are significant differences in how they are represented in the computer models. There are few observations from which a judgement can be made as to the more realistic specification. The uncertainties are significant contributors to the wide range of the surface air temperature predictions from the different models.

These criticisms of the current state of computer model development should not be interpreted as reason for abandoning the technology. In the early 1970s computer models for weather forecasting were similarly the subjects of criticism, but within two decades their skill had advanced to a stage where they consistently outperformed all other techniques. A similar trend of improvement is expected with computer models of the climate system. However progress is likely to be slower because of the added complexity of the complete climate system and the lack of data relating to important climate processes, especially of the ocean.

We return to the question of whether there has been an unfounded rush to judgement over human influence on climate change.

Certainly, the concern about potential for climate change is not unfounded. Extreme events within the variability of the climate system are, by far, the largest cause of natural disasters worldwide each year. Prolonged drought, episodes of flood, tropical cyclones, hailstorms, tornadoes and gales are not random events. These phenomena come about because of the confluence of conditions that increase the potential for the particular event to occur. In a characteristic way we see this confluence during El Niño events, when there are recurring periods of regional drought or flooding. There are other climatic patterns linked to the changing ocean circulation for which an understanding is now being developed. It is important that events such as decadal droughts and multi-decadal warming are understood within the context of complex internal variability of the climate system and possibly forcing from external factors.

There is a need to better understand the climate system, including how it has varied in the past and the processes that contribute to its variability and extremes. Continued support for the international climate-related research programmes is essential to develop the skill for early warning of dangerous events and for the development of strategies to mitigate the impacts that regularly contribute to human suffering and community loss.

The evidence advanced by the IPCC, that human activity will cause dangerous interference with the climate system, is illusory. The assertion that recent warming is unprecedented relies largely on a reconstruction of surface air temperatures covering the last millennium that is fraught with problems of methodology and contentious data. The assertion that the climate system has only limited internal variability is an outcome of constraints on computer models that are necessary to ensure their stability. These assertions ignore a large body of palaeoclimate evidence that points to climate having changed abruptly and relatively frequently in the past. The computer models on which predictions of human interference are based are still at a rudimentary stage of development with major uncertainties about the specification of critical processes. The theory of 'greenhouse climate change' is conceptually simple, seemingly plausible, but deficient in its consideration and treatment of complex climate processes. Importantly, there is no evidence that a reduction in global anthropogenic emissions of greenhouse gases will reduce the incidence of dangerous climate extremes, or reduce the human suffering and community loss that accompany them.